FAMOUS CHEMISTS

THE MEN AND THEIR WORK

SIR WILLIAM AUGUSTUS TILDEN

FAMOUS CHEMISTS AND THEIR LIFE SPAN

Robert Boyle	-	1627-1691
Joseph Black	-	1728-1799
Joseph Priestley	-	1733-1804
Henry Cavendish	-	1731-1810
Carl Wilhelm Scheele	-	1742-1786
Antoine Lavoisier	-	1743-1794
Humphry Davy	-	1778-1829
John Dalton	-	1766-1844
Gay-Lussac	-	1778-1850
Joseph Proust	-	1755-1826
Jacob Berzelius	-	1779-1848
Michael Faraday	-	1791-1867
Amedeo Avogadro	-	1776-1856
Stanislao Cannizzaro	-	1826-1910

FAMOUS CHEMISTS AND THEIR LIFE SPAN

Justus von Liebig	-	1803-1873
Jean-Baptiste Dumas	-	1800-1884
Edward Frankland	-	1820-1899
Alexander Williamson	-	1824-1904
Dmitri Mendeleeff	-	1834-1907
William Crookes	-	1832-1919
William Ramsay	-	1852-1916

FAMOUS CHEMISTS

THE MEN AND THEIR WORK

BY

SIR WILLIAM A. TILDEN, FRS

D.Sc., LL.D., Sc.D.

Professor Emeritus of Chemistry in the Imperial College
of Science and Technology.

AUTHOR OF
" CHEMICAL DISCOVERY AND INVENTION IN THE TWENTIETH CENTURY "

DIAMOND BOOKS™

www.diamondbooks.ca

TORONTO, CANADA – 2017

DIAMOND BOOKS - CANADA

DIAMOND BOOKS

Toronto, ON, CANADA
http://www.diamondbooks.ca

BIBLIOGRAPHIC INFORMATION

' **FAMOUS CHEMISTS - THE MEN AND THEIR WORK** '
was first published in 1921,
by: GEORGE ROUTLEDGE & SONS, LTD.,
LONDON.

PUBLISHED IN CANADA

Published in Canada by DIAMOND BOOKS - CANADA, an imprint of DIAMOND PUBLISHERS - http://www.diamondpublishers.com

REPUBLISHED EDITION: August, 2017.

PAPERBACK EDITION : ISBN: 978-1-988942-20-9
E-BOOK EDITION : ISBN: 978-1-988942-21-6

PRINTED IN CANADA

FAMOUS CHEMISTS
THE MEN AND THEIR WORK

BY

SIR WILLIAM A. TILDEN, F.R.S.
D.Sc., LL.D., Sc.D.
Professor Emeritus of Chemistry in the Imperial College of Science and Technology

AUTHOR OF
"CHEMICAL DISCOVERY AND INVENTION IN THE TWENTIETH CENTURY"

LONDON
1921

PREFACE

THIS book has no pretension to the character of another history of Chemistry. We already possess many histories, of which some, like Kopp's well-known *Geschichte* and Thomson's ***History,*** relate to times now long gone by and have not been continued to the present; others relate to particular periods and are either too superficial to be exact or furnish merely a catalogue of names. There are also a few biographical and autobiographical works, each of which in its way is interesting and valuable. Such, for example, are Henry's *Life of Dalton*, Wilson's *Life of Cavendish*, the *Life of Davy* by his brother John, Bence-Jones' *Life of Faraday*, Frankland's *Autobiography*, Berzelius' *Autobiography*, *Lavoisier* by Grimaux, *Robert Boyle* by Flora Masson, *Ramsay* by myself, *Life and Letters of Black* by Ramsay, *Pasteur* by Valery Radot, of which an English translation by Mrs. Devonshire has recently been issued. But many others have not been written, except in comparatively short obituary notices issued by various scientific societies; in this category the name of A. W. Williamson is conspicuous. Thorpe's *Essays in Historical Chemistry* is a well-known volume, but the biographical sketches, admirable as they are, are independent of one another and have no connecting link.

The origin of the present work was a suggestion from a friend which harmonised with my own experience. Many times in conversation I have been painfully struck by the general ignorance concerning the personal history and doings of men who have created epochs in their respective sciences. Among those who have supplied indispensable links in the chain of events which has culminated in the present state of knowledge in chemistry there are but few whose names are known even among well-educated people. And yet the story of their lives is full of interest and instruction for all who care about the progress of science.

The present volume is an attempt to sketch, in a style suitable for general reading, the lives of some of the most prominent chemists of the past. As it is obviously impossible for one person to deal with them all, the first great difficulty in such a project is to make the selection. After much reflection, the guiding principle adopted is the evolution of the Atomic Theory from the point of view of the Chemist. The very numerous modern applications of chemistry have been explained in a popular form in *Chemical Discovery and Invention in the Twentieth Century* (Routledge) by the same author. Present views of the constitution of matter have resulted from the establishment, step by step, of the idea of the chemical element, the isolation of the chief elements, the establishment of the quantitative laws of chemical combination, the application of the atomic doctrine to the explanation of chemical phenomena, the application of the idea of valency to the constitution and structure of the molecule and the study of the properties of atoms. In the following pages it is the men who have taken the leading part in each step forward along this line whose story is told.

The first question is, Who is to be included?

A crowd of names comes at once into the mind, and each man seems entitled to be called a "great chemist"; but of those who have not been included it will be found, I believe, that, however important their contribution to science, their discoveries have not been *indispensable* to progress along the line indicated. No disrespect is intended to any one of them, but the necessity for limitation imposes a restraint which it would be easier to disregard than to comply with. A case in point is that of Liebig, whose name is included, although it is not easy to point to any direct assistance derivable from his researches. His work in connection with the theory of compound radicals may be accepted as an excuse, but perhaps the immense influence of his teaching at Giessen may be regarded as a better title. Something similar may be said of Cannizzaro. There can be no question as to the services rendered by Dulong, Mitscherlich, Wöhler, Graham, Laurent, Gerhardt, Stas, Kekulé, Hofmann, Wurtz, and others, but they have not been included because their researches led to results which were contributory but not essential to the establishment of the Atomic doctrine.

As to the story of each life, original sources of information have been sought, and statements of fact verified. Of course the general colour imparted to each is the outcome of the writer's own view of the character and career under consideration, and so far the responsibility is his.

The author's cordial thanks are due to the Council of the Chemical Society for permission to reproduce the portraits of Cannizzaro and Mendeléeff, as well as to several friends for assistance or information. Among these must be specially mentioned Professor Svante Arrhenius of Stockholm, Dr. Im. Björkhagen, Professor of Swedish in University College, London, Professor Albin Haller of Paris, and Mrs. A. W. Williamson.

He is further deeply indebted to Dr. Martha Annie Whiteley, of the Imperial College of Science and Technology, for reading the whole of the proofs.

This opportunity must also be used for reporting a fact of which the author has only recently become aware. Robert Boyle is said to have been buried in the chancel of St. Martins-in-the-Fields. This statement is made in Birch's *Life of Boyle* and has been repeated in other biographies. The church referred to was, however, pulled down in 1721 and was replaced by the existing church. In the destruction of the old church the monuments and the remains buried there were all removed, but there is no complete record of what became of them. The enquiry is being pursued, but up to the present with no satisfactory result.

W. A. T.

NORTHWOOD, *March*, 1921.

PRINCIPAL AUTHORITIES

Biographisch-literarisches Handwösterbuch, von J. C. Poggendorff. Leipzig. 1863. With supplements.

Works of the Honourable Robert Boyle, with Life, by Thomas Birch. Five volumes, folio. London. 1744.

Robert Boyle, a Biography, by Flora Masson. Constable. London. 1914.

Lectures on the Elements of Chemistry, by Joseph Black. Two volumes. Edinburgh. 1803.

Life and Letters of Joseph Black, by Sir William Ramsay. Constable. London. 1918.

The Electrical Researches of the Hon. Henry Cavendish. Edited from the original MS. in possession of the Duke of Devonshire by J. Clerk-Maxwell, F.R.S.. Cambridge. 1879.

Life of the Hon. Henry Cavendish, by George Wilson. Cavendish Society. London. 1851.

History and Present State of Electricity, by Joseph Priestley. London. 1767.

Experiments and Observations on Different Kinds of Air, by Joseph Priestley. Three volumes. London. 1774-77.

Priestley. Album containing portraits, etc., collected by James Yates, 1864. In the library of the Royal Society.

Carl W. Scheele: Nachgelassene Briefe und Aufziehungen heraus, von A. E. Nordenskiöld. Stockholm. 1892.

Antoine Laurent Lavoisier: Œuvres publiées par les soins du Ministre de l'Instruction Publique et des Cultes. Tomes 4. Paris. 1864-68.

Traité Élémentaire de Chimie. Paris. 1789.

Lavoisier, 1743-1794, *par* Edouard Grimaux. Paris. 1888.

Sir Humphry Davy. Collected works edited by John Davy, M.D. Nine volumes. London. 1839-40.

Memoir of the Life of Sir Humphry Davy, by John Davy, M.D. Two volumes. 1836.

Humphry Davy, Poet and Philosopher, by T. E. Thorpe. Cassell. 1896.

Life and Letters of Michael Faraday, by H. Bence-Jones. Two volumes. 1870.

The Royal Institution, its Founder and First Professors, by H. Bence-Jones. 1871.

Opere Scelte di Amedeo Avogadro, pub. dalla R. Accademia delle Scienze di Torino. 1911.

Jean Baptiste André Dumas, by A. W. HOFMANN. Extra number of *Nature*, vol. xxi., pp. i. to xl., with portrait. Not indexed in the volume.

French translation, with portrait, in the *Moniteur Scientifique Quesneville*. Paris. April, 1880.

Joseph Louis Proust. Catalogue du Musée Centennial de la Classe 87. Paris.

Jakob Berzelius, Selbstbiographische-Aufzeichnungen. Herausgegeben von H. G. SODERBAUM, nach der wörtlichen Uebersetzung von Emilie Wöhler. Kahlbaum's Monographieen, vii. Heft. Leipzig. 1903.

History of the Medico-Chirurgical Institute of Stockholm. Part iii. Stockholm. 1910. (Swedish. No translation.) The only copy in England is in the library of the Royal Society of Medicine.

HENRY'S *Life of Dalton*. Cavendish Society. London. 1854.

Gay-Lussac : Proceedings of the Royal Society of London. Vol. v. Nov., 1850.

Cannizzaro, S. : Memorial Lecture to the Chemical Society, 1912. Trans. Chem. Soc.

Hofmann's Faraday Lecture to the Chemical Society. 1875. Trans. Chem. Soc.

Liebig, J. : Lecture at Oxford Summer Meeting, 1911. Reported fully in *Nature*.

Dumas, J. B. A. : Obituary notice by A. W. HOFMANN. *Nature*, Feb. 6., 1880.

Sketches from the Life of Edward Frankland. Privately printed, 1902

Obituary Notices of A. W. Williamson. Proceedings of Royal Society. Vol. lxxviii., 1907 : Trans. Chem. Soc., 1905.

Mendeléeff, D. I. : Memorial Lecture to the Chemical Society. Trans. Chem. Soc. 1909.

Crookes, W. : Obituary Notice in Proceedings of Royal Society. 1919.

Ramsay, Sir William : Memorials of His Life and Work, by SIR WM. TILDEN. Macmillan. London. 1918.

CONTENTS

LIST OF ILLUSTRATIONS

Group I

ROBERT BOYLE

THE FATHER OF CHEMISTRY

(1627—1691)

CHAPTER I

BOYLE

CHEMISTRY as a branch of physical science belongs to quite modern times. It is true that many arts in which chemical facts and principles are involved were practised in remote antiquity. For example, glass, which consists of an artificial mixture of silicates made by melting together some kind of alkali with sand or flint was known long before Greek or Roman times, and many specimens of elegant glass vessels of those periods are to be seen in museums in which antiquities are preserved. The bottles used for water or wine mentioned in the Old Testament were doubtless made of pottery or skins of animals, but in Revelations (iv. 6) the description of "the sea of glass like unto crystal" shows that in those days, the first century of the Christian era, the appearance of glass was familiar. Soap again for use in washing was known long ago, though when actually in common use for this purpose is uncertain. In the excavations of Pompeii a soap-boiling establishment was discovered, so that it was known to the Romans. However, wood ashes, and especially the ash of several plants growing near the sea, were in early times the only substances in general use for cleansing purposes.

The art of dyeing has been practised from remote antiquity, and evidence of the high degree of success attained in the operation of fixing colouring matters and producing patterns on linen cloth is afforded by the mummy cloths of Egypt, several thousand years old, which are to be seen in the British Museum and elsewhere.

The extraction of several metals, such as lead, copper, iron, tin, gold, silver from their ores, has also been known from very early times, and the chemical action of fuel in reducing metals from their mineral forms has been utilised, though of course the true explanation of the change was quite unknown. Later in time the alchemists of the fourteenth, fifteenth and sixteenth centuries learnt by mere experiment to make a number of definite chemical compounds, such as nitric, sulphuric and muriatic acids, and they recognised the power of dissolving gold possessed by a mixture of nitric and muriatic acids, hence named *aqua regia*, gold being referred to in alchemical writings as either *sol*, the sun, or *rex metallorum*, the king of metals. Roger Bacon is credited with the invention of gunpowder in the thirteenth century, though it is probable that a similar mixture of nitre, sulphur and charcoal had been known to the Chinese long before that time. Phosphorus is another substance interesting to the chemist which was first obtained by an alchemist, Brandt of Hamburg, in the seventeenth century.

It is certain, therefore, that many chemical elements and definite compounds were known and several important arts dependent on chemical principles were practised long ago. The knowledge of these and similar operations was acquired partly by accident, partly by crude experiments, and does not constitute science. Knowledge only becomes science when accurately established facts are linked together by some explanation, which is referred to as a "hypothesis" or a "theory," according to the solidity of the foundation on which it is believed to rest. It is unnecessary to do more than mention the most ancient doctrines concerning the constitution of matter, for they contributed in no way to the advancement of science, but rather retarded it. The so-called elements, *fire*, *air*, *earth* and *water*, were probably regarded by Aristotle as merely symbolical of the diverse properties of matter, but by most philosophical writers, as well as in popular parlance, they continued to be referred to as the fundamental principles of all things through long centuries down to the times of the alchemists. Evidence in favour of this idea was supposed to be derived from observations of common phenomena. For example, it would be pointed out that where wood burns, there is the display of fire, and an evolution of smoke (air), water escapes as steam, and finally there is a residue of ash, which is manifestly of the nature of earth. The time of exact observation and accurate experiment, however, did not arrive even with the alchemists. Their laborious operations were directed chiefly to the discovery of a universal medicine, the property of which would be to cure all human diseases, and to transmute the base metals, such as lead and copper, into the noble metals, silver and gold. A number of useful discoveries were made by the alchemists, but their experiments being chiefly of a qualitative character, and little importance being attached to the weights of either materials or products, most of their observations were of little value. In the endeavour to find an explanation of the results of their experiments they were led to substitute for the four elements of Aristotle a new system of three first principles, *tria prima*, which were named *salt*, *sulphur* and *mercury*. Apparently, however, none of the writers belonging to the alchemical period down to near the end of the seventeenth century had any clear idea of what was to be understood by these terms. The salt or sulphur extracted from the material operated on, whether of mineral, vegetable or animal origin, did not always present the same characters and was itself manifestly a mixture of more than one thing. Consequently the idea of an *element* as something simple or uncompounded and incapable of analysis into two or more distinct ingredients had not taken definite shape. The fundamental idea necessary to the science of chemistry is a clear definition of the terms *element* and *compound*. If A is united to B the result is indisputably compound. But whether A is to be called an element depends on whether it is possible to take anything out of it which is different from the original A, leaving a residue which is also different from A. Thus there can be no doubt that iron rust consists of a compound, because by suitable experiment both metallic iron and oxygen can be extracted from it, while iron and oxygen are regarded as simple or elementary kinds of matter because from them can be produced only iron and oxygen respectively and nothing else. This conception involves another which, though commonly enough accepted, is usually neglected or unexpressed in popular language, and that is that every substance has fixed properties of its own by which it is identified. How, for example, is water recognised? The answer is that it is a tasteless, incombustible liquid which sets into a crystalline solid (ice) at 0° C. (32° F.) and boils under atmospheric pressure at 100°C. (212° F.), and when seen in small quantities appears colourless. There are no two kinds of water; rain, river, spring and sea waters owe their differences of taste, colour and other properties to the presence of gases, salts and other substances added to the water by natural operations.

The philosophers of ancient times settled everything by conjecture. They supposed that the earth was flat, and that the heavenly bodies existed for its benefit and moved round it in orderly procession. Similarly they supposed that all material bodies were made of fine particles or atoms, as already mentioned, of certain primal stuffs, symbolised by the four elements, fire, air, earth and water, while some even conceived the idea that there was one element only, namely, water.

Science, however, rests only on observation and experiment, and whatever hypothesis or theory is adopted in order to classify or explain the phenomena observed it ought to be received only as a temporary expedient necessary for the assistance of the mind, and liable to be superseded when the state of knowledge has advanced far enough. Hence in respect to the science of chemistry, while there can be no difference of opinion concerning materials which are known to be compounded or mixed, the idea attached to the word *element* may undergo modification in course of time, and whether the number of substances to be included under this term is limited, and how they are to be recognised as *primal* remains to a certain extent an open question. But so long as confusion in language concerning the subject remained there must be confusion of thought, and although many useful facts were known, as already explained, no *science* of chemistry could exist.

The first writer to bring light into the darkness previously existing was therefore rightly called the "Father of Chemistry." This was the famous English experimental philosopher, the

Honourable Robert Boyle.

Son of Richard, the first, the "Great," Earl of Cork, Robert was born on January 26th, 1627, in the reign of Charles I. The circumstances of his life illustrate in an interesting way the frequent coincidence of intellectual activity with political or revolutionary disturbance in a nation. For Boyle was a witness of the destruction of the British Monarchy and its restoration, violent parliamentary oscillations and repeated changes in the national religion, the Great Plague and the Great Fire of London. Among Boyle's contemporaries may be counted not only Isaac Newton, but more than one philosophical and historical writer of the first rank, John Locke and Clarendon, for example; the architect Christopher Wren—the poets Milton, Dryden, Butler, and others. John Bunyan was two years younger, and William Penn, the founder of Pennsylvania, born in 1644, was to lead the defence of the ever-persecuted Quaker Friends. The diarist, Samuel Pepys, was six years Boyle's junior, and afterwards associated with him in the Royal Society, as also was Evelyn, an intimate friend of Boyle, who survived him some fifteen years.

An account of Robert Boyle during his minority was left by Boyle himself under the pseudonym *Philaretus*, and from this autobiography, though never completed, several incidents of his early life may be extracted. As soon as possible after his birth, "when Philaretus was able without danger to support the incommodities of a remove, his father, who had a perfect aversion for their fondness who use to breed their children as if they were made of butter or of sugar, sends him away from home and commits him to the care of a country nurse, who by early inuring him, by slow degrees, to a coarse but cleanly diet and to the usual passions of the air, gave him so vigorous a complexion that both hardships were made easy to him by custom, and the delights of conveniences and ease were endeared to him by their rarity."

Robyn, as he was called by his father, did, however, "ever reckon it amongst the chief misfortunes of his life that he did never know her that gave it him," for his mother died in 1630 when he was about four years old. All his brothers and sisters were put out to nurse in the same way. There was a very large family, and many of their descendants are living at the present day.

Roger Boyle (the first), born at Youghal in 1606, was sent to school in England when he was seven years old. He stayed the greater part of the time with relatives of his mother, at Deptford, and there after a short illness he died. In those days illness was a terrible misfortune to befall a child, for if the disorder itself did not prove fatal the ministrations of the doctor were usually well calculated to bring about the same result. The child was buried in Deptford Parish church. At the time of his death there were four daughters, and in 1612 a son Richard was born. He became Lord Dungarvan and succeeded his father as second Earl of Cork, and in 1663 was created Earl of Burlington. A fifth daughter, Katharine, was born, and at the age of sixteen became the wife of Richard Jones, third Viscount Ranelagh; she was in after life the dearest friend and companion of her young brother Robert, the philosopher. Robert was the seventh son and fourteenth child out of a total of fifteen born to the Earl and Countess of Cork. At the time of his birth his two eldest sisters were already married, and the others passed the same way as soon as they were old enough, the age of fourteen being no uncommon period in those days.

During the years which followed Robert's birth at Lismore Castle there was much journeying to and from Dublin and London, and on one of these excursions from Dublin to Lismore the family coach in which the boy was placed was overturned in crossing a stream and he narrowly escaped drowning. The boy was now being taught under private tutors to write and read and to speak a little Latin and French, and by the time he was nine years of age he was sent, together with his brother Frank, who was twelve, "to be schooled and bred at Eaton." In those days the journey from the Irish home to school was no easy matter. On land they rode on horseback, and to cross the sea it was often necessary to wait for a wind. On September 9, 1635, the boys, under the charge of two servants, set out from Lismore, and after waiting at Youghal, detained by the weather, they travelled by way of Bristol and ultimately arrived at Eton on October 2nd. Robert was a somewhat delicate, quiet boy with a slight stammer, but reported to his father as preferring "learninge afore all other virtues and pleasures," and by no means fond of games, as his brother was.

The doings of the two boys at Eton were reported to their father chiefly by their servant Carew, who had been one of their escort from home. Sir Henry Wotton, the Provost, had received them very kindly, had provided them with a bed till their own quarters were ready, and had assigned them to the care of Mr. John Harrison, the "chief schoolmaster" or Rector of the College.

The boys dined in hall with the other boarders, and at night they supped in their own rooms or in those of some of their companions, among whom were the sons of the Earl of Northampton and those of the Earl of Peterborough. Apparently they enjoyed but little indulgence in the way of food or hours. "They are upp every morning at half an hour afore 6, and so to scoole to prayers." It may be surmised that the discipline at Eton, which was so severe in Roger Ascham's time as to have furnished the occasion for the composition of his "Scholemaster" about sixty years before, had been somewhat modified, at any rate in favour of the sons of noblemen. Possibly the easy-going disposition of the Provost may have contributed to some relaxation. And Robert in his own recollections of Eton sets down in *Philaretus* a full recognition of his indebtedness to the master John Harrison, in whose house he lived and under whose gentle and indulgent supervision he spent three happy years. Other recollections of his schooldays include several accidents, each of which came near to putting an end to his career before it was seriously begun. The most extraordinary of these accidents was the falling of a great part of the wall of their bedroom in Harrison's house. Robyn in the big four-post bed was nearly suffocated with dust, but wrapping his head in the sheet escaped all but inconvenience. Frank, however, talking with some other boys round the fire, was nearly crushed by the falling ceiling and the furniture from the room above.

Another accident was due to the mistake of the apothecary, who gave him a dose out of the wrong bottle. Fortunately, being more than usually repulsive, his stomach failed to retain the potion. "This accident," says *Philaretus*, "made him long after apprehend more from the physicians than the disease, and was possibly the occasion that made him afterwards so inquisitively apply himself to the study of physick, that he might have the less need of them that profess it."

In November, 1638, the boys were removed from Eton and carried to their father's manor at Stalbridge in Dorsetshire. The house and lands had been purchased by the Earl about two years earlier. This investment was probably part of a plan to enable him to have a residence nearer to the King and Court than the castle at Lismore on the other side of the water. These were days full of trouble for the old man: the contests with Wentworth (afterwards Earl of Strafford), the Lord Deputy in Ireland, and, what was worse, the action of the Star Chamber in reference to his title to his lands in Ireland. This in the end cost him £15,000 by way of redemption, and must have severely taxed the resources by which the great establishment at Lismore was kept going.

At Stalbridge the boys were soon put under the charge of the parson as boarders and pupils, undisturbed for the time by the beginnings of civil war in which their father and several of their brothers were taking part on the Royalist side. It was not long, however, before, in accordance with the custom of those days, the elder boy, Frank, was engaged to a young maid-of-honour to the Queen; and in October, 1639, was married to Elizabeth Killigrew with great ceremony in the chapel at Whitehall. The bride was then fourteen years of age and the bridegroom about sixteen. Marriage, however, did not prevent his being sent a few days later with his brother Robert under the charge of a tutor, M. Marcombes, of Geneva, to travel on the Continent.

Marcombes had already, during two years on the Continent, had charge of Robyn's two elder brothers. One of these was Roger the second, born in 1621, afterwards Lord Broghill and first Earl of Orrery. The other was Lewis, Lord Kynalmeaky, born in 1619, and killed at the battle at Liscarrol in 1642. Marcombes had been very justly recommended to the Earl by Sir Henry Wotton, and in the five years which followed he showed himself not only a good teacher but a judicious and friendly companion and guide for the two boys, Frank and Robyn.

The whole party of five persons, including two servants, sailed from Rye, and after a rough passage they landed at Dieppe and pushed on to Paris. Here, joining with other travellers, they made up a party of twenty horsemen, all suitably armed with swords and pistols and, journeying on, in nine days they reached Lyons. After a stay for a time in Lyons the boys and their tutor crossed the mountains of Savoy, and in three days' time reached Geneva, where M. Marcombes resided with his wife and family. There for about two years the young brothers remained. They learned not only to speak the French tongue fluently but were instructed in the usual school subjects, which then included fencing, dancing and fortification. Above all they were under the influence of Calvinistic protestantism, by which they were protected from intercourse with Roman Catholic priests or laity. This undoubtedly assisted in giving a direction to Robert's serious character, which retained its influence throughout his life.

Robert was reported by M. Marcombes always studiously inclined, and preferring the reading of romance or history to such games as "tennisse." They were well fed and well cared for, and *Philaretus* speaks in very high terms of M. Marcombes, who indeed appears to have played the part of a generous friend at the time, a little later, when Robert found himself in a dilemma.

While at Geneva "an accident which he always used to mention as the considerablest of his whole life" occurred to Robert Boyle. And as it gives a view of one side of his character a short account of it must not be omitted. The whole story is told by *Philaretus*:

"To frame a right apprehension of this you must understand that though his inclinations were ever virtuous, and his life free from scandal and inoffensive, yet had the piety he was master of already so diverted him from aspiring unto more, that Christ, who long had lain asleep in his conscience (as He once did in the ship) must now, as then, be waked by a storm." He found himself, in fact, at midnight in the midst of one of those great thunderstorms which happen once or twice in every summer season among the mountains. The storm would appear to have been one of unusual violence and in his terror at what appeared to him the approach of the day of judgment, the boy was led to "resolve and vow that if his fears were that night disappointed all his further additions to his life should be more religiously and watchfully employed. The morning came and a serener cloudless sky returned, when he ratified his determination so solemnly that from that day he dated his conversion; renewing now he was past danger the vow he had made whilst he believed himself to be in it."

Nothing on record in regard to his later life would seem to suggest that the vow thus made and confirmed had lost anything of that controlling and directing influence which manifestly played a part in all the days of his youth.

In the spring of 1641 they bought horses and, longing for a holiday, the tutor and the boys set out for an excursion into the neighbouring country. On their return to Geneva they found letters from the Earl giving them leave, evidently reluctantly, to travel in Italy. The preparation for such a journey was no light matter. Madame prepared for them new linen and M. Marcombes bought for them new suits of clothes. The father's anxieties for their safety in a strange country were mingled with ambition for their instruction and advancement. So after due preparation they set out in September, 1641, all well-horsed, and again they found themselves in the mountains. *Philaretus* tells so much of their journey that their route can be followed across the Grisons into the Val Tellina and so on to Bergamo and the plains of Lombardy, ultimately reaching Florence. Here they sold their horses and settled down for the winter. Both the boys saw much which both amazed and amused them; Frank attracted by the gaiety of the carnival and the balls, while Robyn, having carried his books with him, often insisted on reading even on the road. In March, 1642, they visited Rome, where they saw the Pope and his Cardinals in chapel, but to escape the heat they soon returned to Florence. They then made their way gradually northward, and in May, 1642, they reached Marseilles, where they expected letters from home. Here they were met by news of serious misfortune, for when, after some days, it arrived, the letter from the Earl told of his great difficulties in regard to money, and though forwarding a remittance for their expenses through his agent in London, it never reached them. Robert was then only fifteen years of age, a boy of sedentary habits, a student, a dreamer. Without funds all they could do then was to separate, Frank, the elder, making the best of his way to Ireland, while Robert returned with M. Marcombes to Geneva. Here he had to remain with the Marcombes family during the hard and troubled times in Ireland in which one of his brothers was killed and his father died. In 1644, with the assistance of Marcombes, Robert got back to London from Geneva, after an absence of nearly five years.

England at this time was in the throes of the civil war, and even families were divided. All Robert Boyle's brothers had been or were in the Royalist army, while his sister, Lady Ranelagh, leaned toward the Parliamentarians, and showed her sympathies when she befriended the poet Milton in the days following the Restoration, when all Cromwellians were in danger and some suffered with their lives.

Philaretus therefore reports that "upon his return to England he went immediately to his beloved sister Catherine, Viscountess Ranelagh, a lady remarkable for her uncommon genius and knowledge."

There was no question of his joining the King's army, or, indeed, of fighting on either side. As soon as possible he went to take possession of his own manor of Stalbridge, left to him by his father. He then proceeded to raise money enough to repay the debt he had incurred to M. Marcombes and, with leave obtained from the Parliament, he visited the Continent in the autumn of 1645.

It was now time for him to settle down and do what was possible to put order into his affairs. Boy though he was, being then barely twenty years of age, he took a serious view of all the affairs of life, not only personal, but political, and especially religious. He carried on by himself something of the studies undertaken under M. Marcombes, to whom he wrote giving an account of what he was doing and asking for information about the "ways of husbandry" practised about Geneva. But beside writing "divers little essays both in prose and verse," a practice to which he was much addicted all his life, he was applying himself to "natural philosophy, mechanics and husbandry according to the principles of our new philosophical college." Later in the letter he invited M. Marcombes to bring to England any good receipts or good books "which will make you extremely welcome to our *invisible college*." Boyle was already admitted to the meetings of the club in London, which was the forerunner of the Royal Society.

The history of the Royal Society has been written so often that it is unnecessary to do more in this place than recall the fact that about the year 1645 "divers worthy persons inquisitive into natural philosophy and other parts of human learning, and particularly of what hath been called the *New Philosophy* or *Experimental Philosophy*," began to meet together in London, sometimes in a private house, sometimes at other convenient places, as in term-time at Gresham College. Their business was to discuss matters included under "Physick, Anatomy, Geometry, Astronomy, Navigation, Staticks, Magneticks, Chymicks, Mechanicks and Natural Experiments, with the state of these studies as then cultivated at home and abroad." About the year 1648 or 1649, some of the company having removed to Oxford, the members of the club remaining in London continued their meetings there, while the residents in Oxford carried on similar meetings, part of the time in the lodgings of Robert Boyle. The Oxford Society became in 1651 the Philosophical Society of Oxford, which continued to hold meetings till 1690, when they ceased. The London Society continued to meet until the year 1658, usually at Gresham College, "till they were scattered by the miserable distractions of that fatal year; till the continuance of their meetings there might have made them run the hazard of the fate of Archemides, for then the place of their meeting was made a quarter for soldiers."

In 1660, at the Restoration, the meetings at Gresham College were revived, and toward the end of the year, the King having been acquainted with their designs and intentions, and expressing his approval, the Society was duly constituted with rules and regulations for their proceedings. In the following year the Society petitioned for a charter of incorporation, and, the charter having been granted, passed the Great Seal on July 15th, 1662. This, therefore, was the beginning of the Royal Society, but a Second Charter granting further privileges was sealed in April, 1663. Robert Boyle was named in the Charter a member of the first Council. The King presented to the Society the silver-gilt mace which is placed on the table at each meeting of the Society to the present day.

Boyle was a frequent attendant at the meetings of the "invisible college" throughout its early days, and as often as he could escape from the solitude of his country retreat at Stalbridge. He had tried to set up a laboratory at his own house but with no great success. This led him to write to Lady Ranelagh a despairing kind of letter. "That great earthen furnace," he wrote, "whose conveying hither has taken up so much of my care, and concerning which I made bold very lately to trouble you, since I last did so, has been brought to my hands crumbled into as many pieces as we into sects; and all the fine experiments and castles in the air I had built upon its safe arrival have felt the fate of its foundation. Well, I see I am not designed to the finding out the philosopher's stone, I have been so unlucky in my first attempts at chemistry. My limbecks, recipients and other glasses have escaped indeed the misfortune of their incendiary, but are now, through the miscarriage of that grand implement of Vulcan, as useless to me as good parts to salvation without the fire of zeal. Seriously, Madam, after all the pains I have taken, and the precautions I have used to prevent this furnace the disasters of its predecessors, to have it transported a thousand miles by land that I may after all this receive it broken is a defeat that nothing could recompense, but that rare lesson it teaches me, how brittle that happiness is that we build upon earth."

At the same time he was suffering the first attacks of pain connected with the renal disorder which haunted him all his life. The years immediately following were not favourable to peaceful pursuits, for the strife continued between King and Parliament which ended in the triumph of the Parliament and the last scene in Whitehall. But away in the west, far from the turmoil in London, Boyle carried on his experiments as actively as his health permitted. In 1632 he was able to revisit Ireland for the purpose of looking after his property there, to see the old house where he was born and to stand before the tomb of his father, the great Earl, at Youghal. During the winter of 1653 he took the opportunity of being in Dublin to gain some knowledge of human anatomy. And, as he wrote to a friend, "that I may not live wholly useless, or altogether a stranger in the Study of Nature, since I want glasses and furnaces to make a chemical analysis of inanimate bodies, I am exercising myself in making anatomical dissections of living animals, wherein I have satisfied myself of the circulation of the blood and the (freshly discovered and hardly discoverable) *receptacula chyli* made by the confluence of the *venæ lacteæ*, and have seen (especially in the dissections of fishes) more of the variety and contrivances of Nature and the majesty and wisdom of her Author than all the books I ever read in my life could give me convincing notions of."

As already mentioned, some of the members of the "invisible college" had migrated to Oxford, and in 1654, to escape from the isolation of Stalbridge, Boyle determined to settle there. He found lodging in the house of Mr. Crosse, an apothecary in the High Street adjoining University College, and there, after meeting for some time at Wadham College, the Invisibles carried on their weekly meetings. Here Boyle, now about eight and twenty years of age, busied himself with studies partly experimental, partly in the endeavour to gain a knowledge of the languages of both the Old and the New Testaments in order that he might read the Scriptures for himself in the original tongues. During these years he was associated with many learned and distinguished men, among whom must be mentioned Robert Hooke, a few years younger than Boyle, but a man of great ability and activity of mind. Hooke became assistant in Boyle's laboratory, and worked with him throughout his investigations concerning what he called the "Spring of the Air."

It must be remembered that the barometer had been invented by Torricelli so recently as 1644, and that in 1650 the famous Magdeburg hemispheres had been constructed by von Guericke. The air gun was also known at this time, and is mentioned by Boyle. It seems therefore remarkable that so many years of experimental labour were necessary before the fact could be established that the air has weight, and in the meantime that attempts to prove it experimentally were received with ridicule.

Boyle's *New Experiments Physico-mechanical touching the*

Spring of the Air and its Effects made for the most part in a new Pneumatical Engine was published in 1660. This book embodies experiments made by Boyle with the assistance of Hooke, and contains a description of the air pump which, in the form it ultimately took, is preserved in the rooms of the Royal Society at Burlington House. By the aid of this engine Boyle showed that air has in it a spring or elastic power whereby when a portion of confined air has been removed by the suction of the piston in his pump the remainder expands so as still to pervade the enclosed space, and a further portion of this thinner air may be removed through the valve by further action of the piston of the pump. He also demonstrated the fall of the mercury in a barometer enclosed in the receiver of his pump, and endeavouring to trace the relation between the height of the mercury and the number of strokes of the piston, he came near to the problem which he ultimately solved and which is associated with his name. He also proved by means of his pump the connection between pressure and the boiling of water.

The philosophers of the eighteenth century were occupied with many metaphysical questions, and among other strange doctrines the maxim that "nature abhors a vacuum" gained very considerable currency on the authority of the famous French philosopher Descartes (1596-1650). It is therefore easy to conceive the astonishment with which Boyle's experiments with his air-pump were witnessed by the numerous visitors, members of the Invisible College and others, who were admitted to the demonstrations. Fortunately the objectors to the theory of the spring of the air were sufficiently persistent to compel Boyle to a formal *Defence of the Doctrine touching the Spring and Weight of the Air* published in 1662. This included an account of the famous *Two new Experiments touching the Measure of the Force of the Spring of Air compressed and dilated*," whereby not only were the objections of Franciscus Linus refuted, but the "Law of Boyle," since known to every student of physical science, was established. It will therefore be interesting to the reader to have before him an essential part of the passage in which the well-known siphon tube is clearly described. (Boyle's Works, vol. i., chap. v., p. 100.)

"We then took," he says, "a long glass tube which by a dexterous hand and the help of a lamp was in such a manner crooked at the bottom that the part turned up was almost parallel to the rest of the tube and the orifice of this shorter leg of the siphon (if I may so call the whole instrument) being hermetically sealed the length of it was divided into inches (each of which was subdivided into eight parts) by a streight list of paper, which containing those divisions was carefully pasted all along it. Then putting in as much quicksilver as served to fill the arch or bended part of the siphon, that the mercury standing in a level might reach in the one leg to the bottom of the divided paper, and just to the same height or horizontal line in the other, we took care by frequently inclining the tube so that the air might freely pass from one leg into the other by the sides of the mercury (we took, I say, care) that the air at last included inthe shorter cylinder should be of the same laxity with the rest of the air about it. This done we began to pour quicksilver into the longer leg of the siphon, which by its weight pressing up that in the shorter leg, did by degrees streighten the included air ; and continuing this pouring in of quicksilver till the air in the shorter leg was by condensation reduced to take up but half the space it possessed (I say possessed, not filled) before ; we cast our eyes upon the longer leg of the glass, on which was likewise pasted a list of paper carefully divided into inches and parts, and we observed, not without delight and satisfaction, that the quicksilver in that longer part of the tube was 29 inches higher than the other. . . . The air in that degree of density and correspondent measure of resistance to which the weight of the incumbent atmosphere had brought it was able to counterbalance and resist the pressure of a mercurial cylinder of about 29 inches as we are taught by the Torricellian experiment ; so here the same air being brought to a degree of density about twice as great as that it had before obtains a spring twice as strong as formerly, as may appear by its being able to sustain or resist a cylinder of 29 inches in the longer tube together with the weight of the atmospherical cylinder that leaned upon those 29 inches of mercury ; and as we just now inferred from the Torricellian experiment was equivalent to them."

In order to measure the spring of rarefied air Boyle then proceeded in the manner familiar to the readers of text books of physics. A narrow tube containing a small portion of air was dipped into the open end of a somewhat wider tube full of quicksilver, and the volume of the included air being measured when the metal within was at the same level as the quicksilver in the wider tube, the narrow tube was lifted till the column of air was expanded to a determinate length. The height of the column of quicksilver in the inner tube was then measured and in this way the relation of pressure to volume could be recorded. Two tables of measures were thus constructed, namely, one containing the results with condensed air, the other those obtained with rarefied air, and a comparison instituted between the observed pressures and those calculated according to the hypothesis that the pressure and volume are in reciprocal proportion. The agreement was found to be within the limits of experimental error and thus the law was established.

Boyle's name is immortalised by his discovery of the law connecting volume and pressure in gases, but though his services to chemistry are equally important it is not so easy to state in a few words the exact nature of his claim to be called "The Father of Chemistry." His work lay in examining the pretensions of the chemists of his time and exposing the mixture of error and imposture contained in the greater part of their writings. His views are set forth in a book under the title *The Sceptical Chymist: or Chymico-physical Doubts and Paradoxes touching the Experiments whereby vulgar spagirists are wont to Endeavour to Evince their Salt, Sulphur and Mercury to be the True Principles of Things.* In this book, the contents of which are in the form of a dialogue, Boyle showed not only that the peripatetic doctrine of the four elements was untenable but that the pretensions of the alchemists in regard to their three principles were founded on bad observation and experiments from which nothing could be rightly concluded. For, as he pointed out, the substances separable from bodies by the action of fire do not only vary in number, but the names, such as sulphur, given to these substances, do not correspond with their properties.

Nor did he hesitate to point to the ignorance of the "chymists" of his time and to complain of "their obscure, ambiguous and almost Ænigmatical Way of expressing what they pretend to Teach that they have no Mind to be understood at all, but by the Sons of Art (as they call them) nor to be understood even by those without Difficulty, and Hazardous Tryalls." Boyle also seems to have had a notion similar to that of modern times as to chemical union taking place between the minute parts or atoms of substances so that from their "coalition may emerge a new body, as really one as either of the corpuscles was before they

An Alchemist in Boyle's time.
From the Picture by *D. Teniers*.

were mingled or, if you please, confounded." He also showed that analysis of compounds cannot be effected solely by the action of fire, but that other agents may be necessary and that in any case the substances resulting are usually not elements in the proper sense of the word, which should be applied only to the products of ultimate analysis. From Boyle's time, therefore, the idea of an element or principle follows from the failure of all attempts to show that a given substance consists of more than one ingredient. The old chemists were therefore justified in placing in the list of chemical elements such substances as lime, magnesia and caustic potash which were not shown to be compounds till more than a century after Boyle's time.

At the time of the publication (1661) of his *Sceptical Chymist*, Boyle was only thirty-five years of age, and the years had not been occupied solely with experimental philosophy. He was a frequent and voluminous writer and he occupied himself not only with such work as *Some Considerations touching the Usefulness of Experimental Natural Philosophy*, but with various essays on religious subjects, and on many of the very diverse questions which attracted the attention of the "Invisibles." A sufficient account of some of these is contained in Bishop Sprat's *History of the Royal Society*.

Once the idea of enquiring into natural objects and phenomena had taken hold of the minds of the philosophers they found themselves bewildered by the illimitable wealth of material, and some of their enquiries were quaint enough. Boyle was the most famous, though naturally one of the most modest, men of the day, and the experiments made and contemplated with his air-pump at Gresham College served not only to excite astonishment, but to direct the thoughts of the men of science to the nature, qualities and phenomena of the air. It is probable that the notion of the materiality of air was established so slowly as to be strange and unfamiliar to many people long after Boyle's time; in fact, until nearly a century later, when "pneumatic chemistry" began to be commonly practised.

His treatise on *Seraphick Love* occupied not only a long time in composition but lay aside for many years before it was ready for publication. Evelyn had read the MS. treatise with delight in 1659, but it was not in print till the following year. However tedious the present-day reader might find it, there seems to have been a romance attaching to this production, for while still a child his father the Earl had set his heart on making a match for Robyn with the Lady Ann Howard, daughter of his friend Lord Howard, of Escrick, and had presented to the young lady a ring as a token and another before he died in 1642. One of these two rings may have been returned to Robyn before her marriage with her cousin Charles Howard, afterwards Earl of Carlisle, in 1645. At any rate a ring which was disposed of in Robert Boyle's will long years afterwards was possibly identical with one of them. The terms of the bequest are as follows:

"I give and bequeath unto my dear Sister, the Lady Katherine, Viscountess Ranelagh, a small ring usually worn by me on my left hand having in it two small diamonds with an emerald in the middle, which ring being held by me, ever since my youth in great esteem, and worn for many years for a particular reason, not unknown to my said sister, the Lady Ranelagh, I do earnestly beseech her, my said sister, to wear it in remembrance of a brother that truly honoured and most dearly loved her."

It appears to be impossible, from the evidence in existence in Boyle's own correspondence or in that of his friends, to determine with certainty to whom this ring originally belonged and the reason for its being treasured and worn by the philosopher during so large a part of his life. That Lady Ann Howard was repeatedly spoken of by the old Earl, his father, as "My Robyn's yonge Mrs." is certain, and that for a time she must have been so regarded by the family is also pretty certain. Lady Ranelagh writing to her brother in, probably, the early part of 1645, refers to the matter in terms which seem clearly to indicate what had been the view of the family. She says, "You are now very near the hour wherein your mistress is, by giving herself to another, to set you at liberty from all the appearances you have put on of being a lover; which though they cost you some pains and use of art, were easier because they were but appearances."

Possibly the explanation is that Robyn had assumed the appearances of the lover in deference to the known wishes of his father. And the pages of *Seraphick Love* may have been the receptacle not so much of disappointment as of those serious views of the obligations of the married state which were so characteristic of him and which he set forth at such length.

There is, however, the statement of his old friend, John

Evelyn, who, writing about Robert Boyle long years afterwards, said: "Tho' amongst all his experiments he never made that of the maried life, yet I have been told he courted a beautifull and ingenious daughter of Carew, Earl of Monmouth, to which is owing the birth of his *Seraphick Love*." Robyn did in the summer of 1648 pay a short visit to the Monmouth family at Moor Park by Rickmansworth, in Hertfordshire. And on the invitation of the Countess he had carried with him a manuscript—it may have been an instalment of *Seraphick Love*—which was completed soon after this time. It is uncertain which of the Countess's daughters was entitled to be called "beautiful and ingenious," but nothing came of the visit and there was no mention of a ring.

In 1662 Boyle received through the influence of friends a further grant of Irish land, which he accepted rather reluctantly, devoting the revenues partly to the relief of the poor and the maintenance of the Protestant religion in Ireland, and partly to the Propagation of the Gospel in New England. In 1665 the increase of the plague in London led to the suspension of the meetings of the Royal Society at Gresham House. Boyle remained at Oxford the greater part of the time, till early in the following year, when the meetings of the Society were resumed. Boyle escaped the plague, but he was frequently ill of his chronic disorder, and it is uncertain whether he witnessed the outbreak of the Great Fire which began on September 3rd. His sister, Lady Ranelagh, assisted in the distribution of his charity among the poor houseless Londoners, while her brother got back to Oxford and calmly pursued his studies.

During 1667-8 the Boyle family were in great anxiety owing to the political changes then going on, which involved Boyle's brother, Lord Orrery, for a time, and in 1669 Boyle left Oxford and took up his residence permanently with his sister, Lady Ranelagh, then a widow, at her house in Pall Mall. Henceforward he led the life of an invalid, and though he had established a laboratory at the back of the house the work was frequently interrupted by attacks of illness which laid him for the time prostrate. This did not prevent the publication of many papers and memoirs in the *Philosophical Transactions of the Royal Society* and elsewhere, and as the acknowledged head of the scientific community in England Boyle was elected President of the Royal Society on November 30, 1680. His health, and probably other considerations, led him, however, to decline the proposed honour, and Sir Christopher Wren was appointed to the vacant chair. Boyle visited no more his manor at Stalbridge, which in his will he left to his brother Frank, Viscount Shannon, but his income must have been considerable from his several estates.

The philosopher and his sister pursued their quiet way during all the frequent and momentous changes going on around them. They received many friends and relatives, and the visits of distinguished foreigners who came to pay their respects to the most eminent man of science in the country. Evelyn, the friend of forty years, would often make his way from Deptford and his pleasant garden at Sayes Court to the house in Pall Mall. Henry Oldenburg, one of the two first secretaries of the Royal Society, had made the acquaintance of Boyle in Oxford about 1656, and had been travelling tutor to Richard Jones, Lord Ranelagh, the husband of Boyle's sister Katherine. Oldenburg had been, from the days of the Invisible College, a correspondent of Boyle's, and reported to him all that was of interest at Gresham College. But he had also done a good deal of work for him in connection with his papers, and when Oldenburg died in 1667, leaving his children ill provided for, it was Boyle and Lady Ranelagh who looked after them.

Meanwhile the world went on. King Charles was succeeded by James, and James by William and Mary, but though the echoes of all the tumult outside must have reached the philosopher, his own precarious state of health, coupled with habit and his own inclinations, tended to sustain the quiet atmosphere of the joint home. Immediately after the Restoration in 1660 the river had been brought through St. James's Park, and many improvements had been introduced into the park itself. These were noticed by Pepys in 1662, and the game of Mall was always popular with the Court and in winter the ice provided for the new sport of skating. Otherwise the Mall was quiet enough as a place of residence. Boyle's health had been from early manhood haunted by the pain and trouble already referred to, but in 1670 he was suddenly seized with a "severe paralytic distemper" from which he recovered only very slowly, and during the twenty years of life still remaining to him he was obliged to withdraw more and more from the meetings of the Royal Society and from those of the East India Company, of which he was a Director, and the

Society for the Propagation of the Gospel, of which he was a Governor.

On December 23rd, 1691, Lady Ranelagh died. She was twelve years older than her illustrious brother, who followed her to the grave seven days later. He died on December 30th, 1691, and both were buried in the chancel of St. Martin-in-the-Fields, facing what became, much later, Trafalgar Square.

Group II

THE PHLOGISTIANS

BLACK (1728-1799) CAVENDISH (1731-1810)
PRIESTLEY (1733-1804) SCHEELE (1742-1786)

CHAPTER II

BLACK

THE creation of the Royal Society and its establishment in a position of influence was accomplished during Boyle's lifetime, and largely by his assistance. That it was the great national event of the reign of Charles II. is proved by the subsequent history of the society during upwards of two hundred and fifty years. Arising out of the slow fermentation proceeding in the minds of men of this period similar movements were taking place in other countries, and the result was the establishment in several of the countries of Europe of Academies having for their object the cultivation of scientific enquiry. But, as already observed, the study of nature presented a field at once so various and extensive, so filled with obscure and perplexing problems as to yield at first little more than confusion. In regard to chemistry the work published during and after Boyle's time for many decades led to few results of importance. The phenomena of fire had naturally attracted attention from the earliest times, but even in Boyle's day ideas as to the cause of combustion were generally erroneous, owing chiefly to the fact that the part played by air was ignored. One exception to this ignorance must be recognised in the observations of a contemporary of Boyle, John Mayow, fellow of All Souls, Oxford. He seems to have discovered experimentally that air contains a *spiritus nitro-aereus* identical with a constituent of saltpetre which combines with metals when calcined and which plays an important part in respiration. His observations, however, were cut short by his early death and do not seem to have attracted the attention they deserved. Progress was delayed in consequence of the fact that chemists had not yet adopted quantitative methods in their work. For no practical purposes except assaying of gold and silver was it customary to

JOSEPH BLACK, M.D.
Professor of Chemistry, Glasgow and Edinburgh.

[*Face p.* 22]

pay any attention to weight, and such observations as those of Jean Rey, who had shown that lead and tin when calcined increase in weight, were ignored. Unfortunately soon after Boyle's time a theory came to be gradually adopted which occupied the field for the greater part of a century. The notion that all combustible substances and metals which were reduced to ashes by calcination contained a common constituent or fire material originated in the writings of a German chemist, Johann Joachim Beecher. On this idea was founded the celebrated theory of *phlogiston* taught by Georg Ernst Stahl, at one time physician to the King of Prussia. The doctrine had apparently one merit in the recognition of the fact that the power of burning was communicable from one body to another. A combustible substance was supposed to consist of two constituents, a *calx* and *phlogiston*.[1] In the process of burning the phlogiston escaped, commonly giving the appearance of flame, and the calx remained behind. In substances like charcoal there was supposed to be a large quantity of phlogiston and little calx; on the other hand, such a metal as lead contained much calx and little phlogiston. By heating together charcoal and lead calx the phlogiston was restored to the calx and metallic lead reappeared. In modern times an analogy has been traced between this phlogistic doctrine and the mutual transference or transformability of potential and kinetic energy. There is, however, but little justification for attempting any such parallel. The adoption of the phlogistic idea was the result of blindness and ignorance, and its strange persistence can only be attributed to the systematic neglect of quantitative methods in the experimental study of chemical phenomena.

To the same neglect of weight and measure may also be attributed, at least in part, the failure of chemists after the time of Boyle and Mayow to recognise the substantiality of the different kinds of air (gases).

The first important step in the direction of reform was taken in the middle of the eighteenth century by JOSEPH BLACK, professor of Chemistry in the Universities of Glasgow and Edinburgh.

Black left a short autobiographical memorandum which,

---

[1] Calx: Latin = ash. Phlogiston: Greek = anything set on fire.

together with a selection of his letters, has been recently published.[1]

The memorandum is as follows: "I was born at Bordeaux, April 16th, 1728. My father was a merchant settled there, but born in Ireland and the son of a citizen of Belfast of Scotch extraction. My father's residence at Bordeaux was in the suburb called the Chartron, and he had also a farm and country house and vineyard on the other side of the river on a hill called Lormont which commands a fine prospect of the river and city. At this villa, and at his house in town, he had sometimes the honour to receive a visit from the Baron de Montesquieu, who had the goodness to favour my father with his friendship and protection. My mother was of the family of Gordon, of Halhead in Aberdeenshire, a branch of which family had also settled in Bordeaux in the mercantile line. I received the first beginning of education from my mother, who taught all her children to read English. At twelve years of age I was sent to Belfast to live with some of my father's relations, and to acquire the rudiments of Latin and Greek. After four years spent there at a private school I was sent to the university of Glasgow, where I attended all the lectures on the languages and philosophy in a regular succession. Being then required by my father to choose a profession or employment I chose that of Medicine, the elements of which I began immediately to study by attending the lectures of the professor of Anatomy and of Dr. Cullen, who was at that time Professor of Medicine at Glasgow. Dr. Cullen began also at this time to give lectures on Chemistry, which had never before been taught in the University of Glasgow, and finding that I might be useful to him in that undertaking he employed me as his assistant in the laboratory, and treated me with the same confidence and friendship and direction in my studies, as if I had been one of his own children. In this situation I lived three years.

"In the early part of my chymical studies, the author whose works made the most agreeable impression on my mind was Markgraaf of Berlin; he contrived and executed his experiments with so much chymical skill that they were uncommonly instructive and satisfactory; and he described them with so much

[1] *The Life and Letters of Joseph Black*, by Sir William Ramsay (Constable. 1918).

modesty and simplicity, avoiding entirely the parade of erudition and self-importance, with which many other authors encumber their works, that I was quite charmed with Markgraaf and said to Dr. Cullen that I would rather be the author of Markgraaf's Essays than of all the chymical works in the library. The celebrated Réaumur's method of writing appeared to me also uncommonly pleasing. After three years spent with Dr. Cullen I came to Edinburgh to finish my education in medicine. Here I attended the lectures of Dr. Munro, senr., and the other medical Professors, until the summer of the year 1754, when I received the degree of Doctor of Medicine, and printed my inaugural Dissertation *De Humore Acido a Cibis Orto et Magnesia Alba.*"

Before proceeding further it is well to remember that at this time and for long afterwards Chemistry, as a subject of study in the universities, was regarded merely as subordinate to medicine, and served to supply materials which were used as remedies in the treatment of disease. Markgraaf (Andreas Sigismund Marggraaf, 1709-1782), mentioned by Black, had been, during all the earlier years of his life, trained and occupied as an apothecary, and this may have given a colour to his chemical work which perhaps had some influence in attracting Black while a student of medicine. Marggraaf was, however, under the influence of the phlogistic doctrine to the end of his life, and though a skilful experimenter who contributed several interesting facts to the history of phosphoric acid, of alum, of sugar and some other compounds, his work had practically no influence on the progress of the science.

Joseph Black was one of a large family, " eight sons and five daughters who all grew up to be men and women and settled in different places." He seems to have been endowed not only with a studious disposition but amiable and courteous manners which endeared him throughout life to many friends.

In 1751 he went to Edinburgh, where he finished his medical studies, and in 1754 he passed the examinations and secured the degree of Doctor of Medicine, having presented the thesis already mentioned in his autobiographical sketch.

On the retirement of Dr. Cullen in 1756 Black was appointed professor of anatomy and lecturer on chemistry in the University of Glasgow, though at that time no more than twenty-eight years of age.

Black seems to have been a man of quiet character, little animated by ambition, and though interested in scientific experiment his enthusiasm seems to have been easily exhausted or not active enough to carry him rapidly forward in any field of enquiry. At any rate, it seems remarkable that Black's discoveries in the years 1754 to 1756, so important as they now appear, should have remained so long without complete and formal publication to the world. The little volume containing an account of his work is dated 1777 under the title "*Experiments upon Magnesia Alba, Quick Lime and other Alcaline Substances,* by Joseph Black, M.D. Professor of Chymistry in the University of Edinburgh." He never adopted the practice of communicating his results to the Royal Society or to one of the scientific academies, and only four treatises appear from first to last under his name. The question which Black set out to investigate was the cause of the difference between the mild and the caustic alkalis, represented, for example, by chalk or by salt of tartar (potassiam carbonate) on the one hand and by lime or by caustic potash on the other. Previously it had been supposed that quick-lime derived its caustic or burning character from the fire in which the original limestone had been burned, and that it was capable of passing on this quality to a solution of mild alkali converting it into caustic alkali; that is, to a form in which alone it possessed the power of converting fat into soap.

The simple experiments of Black disposed of this assumption and provided the true explanation of the change by merely tracing the alterations of weight involved and recognising the part played by a constituent of the atmosphere to which Black gave the name "fixed air." Having heated a quantity of common magnesia to redness and obtained only a small quantity of faintly ammoniacal water, he remarks (p. 29) "even from this imperfect experiment it is evident that of the volatile parts contained in that powder a small proportion only is water; the rest cannot, it seems, be retained in vessels under a visible form. Chemists have often observed, in their distillations, that part of a body has vanished from their senses, notwithstanding the utmost care to retain it; and they have always found, upon further enquiry, that subtile part to be air, which, having been imprisoned in the body, under a solid form was set free and rendered fluid and elastic by the fire. We may therefore safely conclude that the volatile matter lost in the calcination of *magnesia* is mostly air, and hence the calcined *magnesia* does not emit air or make an effervescence when mixed with acids."

Then he shows that the air driven out by calcination of magnesia can be restored to it by an alkali. "Two drams of *magnesia* were calcined in a crucible, in the manner described above, and thus reduced to two scruples and twelve grains. This calcined *magnesia* was dissolved in a sufficient quantity of spirit of vitriol, and then again separated from the acid by the addition of an alkali, of which a large quantity is necessary for this purpose. The magnesia, being very well washed and dried, weighed one dram and fifty grains. It effervesced violently, or emitted a large quantity of air, when thrown into acids; formed a red powder when mixed with a solution of sublimate; separated the calcarious earths from an acid and sweetened lime water: and had thus recovered all those properties which it had but just now lost by calcination. Nor had it only recovered its original properties, but acquired besides an addition of weight nearly equal to what had been lost in the fire; and as it is found to effervesce with acids, part of the addition must certainly be air. This air seems to have been furnished by the alkali from which it was separated by the acid, for Dr. Hales has clearly proved that alkaline salts contain a large quantity of fixed air, which they emit in great abundance when joined to a pure acid. In the present case the alkali is really joined to an acid, but without any visible emission of air; and yet the air is not retained in it; For the neutral salt, into which it is converted, is the same in quantity and in every other respect, as if the acid employed had not been previously saturated with *magnesia*, but offered to the alkali in its pure state, and had driven the air out of it in their conflict. It seems therefore evident, that the air was forced from the alkali by the acid and lodged itself in the *magnesia*."

So that Black's explanation of the interaction between the magnesium salt and the alkali was almost exactly that which would have been given in more modern times. Some readers may be disposed to wonder that so much importance is attached to these simple observations, but it must be remembered that there existed in Black's day very little exact knowledge concerning chemical phenomena generally. Thus it can hardly be said that any substance was known in a state of purity, that is, unmixed with more or less of other matters, inasmuch as there was no criterion of purity. Potash and soda were confused together under the common name of *alkali* and it had only been quite recently noticed that the alkali from wood ashes presented some characters in which it differed from the alkali derivable from common salt. It was not known that they were compound substances and they were indeed ranked as elementary in accordance with Boyle's teaching. Black succeeded in showing that the mild form of alkali was a compound being made up of caustic alkali and fixed air. It is unnecessary to pursue this subject further, but it must not be forgotten that to Black science owes another very important step in advance.

In Black's day, and long afterwards, heat was supposed to be a kind of matter, and even after the discoveries by Count Rumford and Sir Humphry Davy in the early part of the following century it was asserted in a well-known text book (Henry's *Chemistry*, 11th edit., 1829) that "the opinion which best explains the phenomena is that which ascribes them to an extremely subtile fluid, of so refined a nature, as to possess no sensible weight and to be capable of insinuating itself between the particles of the most dense and solid bodies. To this fluid as well as to the sensation which it excites the term *heat* was formerly applied. . . . The term *caloric*, first proposed by Lavoisier, is now frequently adopted to denote the *cause of* heat." Black seems to have held the opinion that heat entered into a kind of chemical combination with ordinary matter, as shown in his discussion of experiments in which he came near to the recognition of what in more modern language would be called Specific Heat. Mercury at 150° having been mixed with an equal measure of water at 100°, the resulting temperature was found to be not 125° but 120°. "The quicksilver therefore is become less warm by 30 degrees, while the water has become warmer by 20 degrees only; and yet the quantity of heat which the water has gained is the very same quantity which the quicksilver has lost. This shows that the same quantity of the matter of heat has more effect in heating quicksilver than in heating an equal measure of water, and therefore that a smaller quantity of it is sufficient for increasing the sensible heat of quicksilver by the same number of degrees." He infers that each substance attracts a quantity of heat peculiar to itself and suggests that "perhaps not any two of them would receive precisely the same quantity, but each, according to its particular capacity, or its particular force of attraction for this matter."

Whatever his views as to the nature of heat, he succeeded in establishing the doctrine of latent heat of fusion and of vaporisation, and he even estimated approximately the amount of latent heat of water and of steam. This work was completed while he remained in the University of Glasgow, but in 1766 Cullen resigned the chair of Chemistry at Edinburgh and Black was appointed to succeed him.

Black appears to have been a careful and conscientious teacher, but the feeble condition of health from which he had long suffered increased with age, and is doubtless sufficient to account for the fact that he declined general practice as a physician in Edinburgh and showed but little activity in pursuing chemical investigation and even in publishing an account of discoveries already made. It is uncertain, for example, how much he knew of the properties of the *fixed air* with which he had so much to do in his investigation of the alkalis and alkaline earths.

His last course of lectures was delivered in the winter of 1796-7, and on November 10th, 1799, in his seventy-first year, he expired quietly in his chair.

England in the middle of the eighteenth century was very different from the country of Robert Boyle, and though less violently disturbed was little more favourably disposed toward learning and the pursuits of science. The Hanoverian succession had secured internal peace, but the state of political parties, the condition of education, of morals and of business can be imagined from the subjects of Hogarth's famous series of pictures, and from the recollection of a few facts which seem significant.

Medicine, the practice of which occupied the attention of Black himself, was based on no secure principles, for even human anatomy was very imperfectly known and there was practically no science of physiology, while the substances employed as remedies were generally useless and often disgusting and dangerous. It was a fair indication of the general state of ignorance prevailing at this period that the virtues of tar-water should have been the subject of more than one disquisition by no less a person than the famous metaphysician George Berkeley, Bishop of Cloyne. As for popular credulity, the belief in witches and astrology was practically universal, while touching for "the king's evil" had been practised down to the time of Queen Anne and was only abandoned because it was supposed that the virtue belonged exclusively to the Stuarts and did not extend to the House of Hanover.

The Universities of Oxford and Cambridge were favourable, according to many writers of the time, as much "to the diffusion of ignorance, idleness, vice and infidelity "[1] as of sound learning and Christian principles, and many stories are extant concerning the habits of the undergraduates and the often farcical proceedings which led to a bachelor's degree. In the Scottish universities discipline appears to have been much more strict, and in Black's time the University of Glasgow seems to have been frequently occupied with controversies concerning the relative powers of the Senate and the faculties. The medical school was fairly active, a considerable body of students gathered round Black, and in 1763 a new chemical laboratory was provided.[2]

Robertson, or Robison, as he chose to call himself, Black's successor in Glasgow and his only biographer, has mentioned the names of his chief friends and associates in Edinburgh. Among these James Hutton the geologist must be mentioned first, as he is represented as "the only person near him to whom Dr. Black imparted every speculation in chemical science and who knew all his literary labours; seldom were the two friends asunder for two days together." Hutton was senior to Black by only two years, and was a valuable and sympathetic critic of Black's ideas, but there were other friends whose society would be appreciated by a man with a taste for scientific pursuits. Adam Smith (1723-1790), author of the *Wealth of Nations*, David Hume (1711-1776), the philosopher and historian, Adam Ferguson (1723-1816), Professor of Philosophy at Edinburgh, John Home (1722-1808), dramatic writer, and Dr. John Roebuck (1718-1794), who had studied chemistry and medicine at Edinburgh and is memorable as having introduced the lead-chamber in connection with manufacture of sulphuric acid, were members of the philosophic and scientific circle centring in the university of which Black was a professor. Among the pupils attending his lectures was Benjamin Rush, a native of Philadelphia, who on returning to his own country became, in the University of Pennsylvania, the first Professor of Chemistry in America.

---

[1] Knox's *Essays*, vol. ii., sec. 29.

[2] *History of the University of Glasgow* (James Coutts. 1909).

By the middle of the eighteenth century alchemistic ideas and the practices of the adepts had become discredited, and chemistry as a department of natural knowledge came to be honestly pursued with the object of penetrating some of the secrets of nature, and it happened, fortunately for the progress of science, that four contemporaries of Black, each possessing special gifts, were tempted into this field of work. Of these four two were Englishmen, namely Joseph Priestley and Henry Cavendish, while Scheele was a Swede and Lavoisier a native of France. All four made discoveries of facts which greatly assisted progress, but it is remarkable that one only possessed the clear vision which enabled him to see through the confusion and obscurity so long spread over all ideas in chemistry by the doctrine of phlogiston.

CHAPTER III

PRIESTLEY

JOSEPH PRIESTLEY was born in 1733 at Fieldhead, a small hamlet about six miles from Leeds. His parents were humble folk, the father, Jonas Priestley, being a cloth-dresser by trade, and when only seven years old he had the misfortune to lose his mother. He was, however, taken charge of by an aunt, and though his early education was somewhat desultory, he seems to have been encouraged in the development of that aptitude for study, and especially for languages, which was one of his characteristics. He was sent to a grammar school, and at sixteen years of age had made considerable progress in Latin, Greek and Hebrew. His aunt, a strong Calvinist, desired him to become a minister of religion, and after some interruptions owing to the state of his health, this course was ultimately adopted and in 1752 he was sent to the theological academy at Daventry, where he remained three years. When he left the academy his first charge was as assistant to the Presbyterian minister at Needham Market. Here his income never exceeded £30 a year, but the congregation disapproved his theological opinions, which had become extraordinarily liberal, and consequently even this failed him. After trying with indifferent success to carry on a school, and applying for various posts, he received an appointment as minister at Nantwich, in Cheshire, where he remained three years. He then became tutor in languages at the recently instituted Warrington Academy, and here his talents and his teaching were recognised as having "added greatly to the celebrity of the institution." During his residence here he was in 1764 admitted to the honorary degree of LL.D. in the University of Edinburgh, and in 1766 he was also elected F.R.S. At

Joseph Priestley.

From the pastel by Mr. Sharples in the National Portrait Gallery.

[*Face p.* 32].

Warrington in 1762 he married Mary, the daughter of Isaac Wilkinson, an iron master at Wrexham, and in the same year he was ordained to the ministry.

After six years in the Academy he accepted an invitation to take charge of Mill Hill Chapel, Leeds. Notwithstanding the check to his health in early youth, Priestley must have been endowed with extraordinary endurance and activity of mind and body, for the long hours given to teaching and study and the discharge of ministerial duties on Sunday did not prevent the production of writings on a great variety of subjects. In 1767, the year of his removal to Leeds, he brought out his *History of Electricity*, and in the following year his *Treatise on Civil Government*. But though his fertility in religious writing, chiefly of controversial character, must always appear surprising, it is as an experimental chemist and especially as the discoverer of oxygen that Priestley will be remembered to the end of time. Already at Nantwich he had become possessed of a few philosophical instruments—such as an air-pump and electrical machine—but he does not appear to have used them except for purposes of instruction.

Priestley's career as a scientific investigator began practically in Leeds, and his place in the history of science rests entirely on his discoveries in connection with gases. These are set forth in the six volumes he published under the title *Experiments and Observations on Different Kinds of Air*, of which the first volume is dated 1774. Here he relates in his usual candid style how he was led in this direction and describes the form of apparatus he employed in his experiments. He says:

> "It was in consequence of living for some time in the neighbourhood of a public brewery that I was induced to make experiments on fixed air, of which there is a large body, ready formed, upon the surface of the fermenting liquor."

The gas which is evolved from the lime-kiln and in the process of fermentation had been observed even so far back as the sixteenth century by Paracelsus, and later by Van Helmont, but its power of entering into union with alkalis was first definitely established, as already related, by Dr. Black. Other kinds of air were also known before Priestley began his experiments. These included the fire-damp of mines as well as the inflammable air obtained by the solution of metals in acids. Previous to Priestley's time various kinds of air were usually collected in bladders, but Priestley introduced the simple method of displacing by the gas a liquid—water or, in some cases, mercury—with which the bottle or other receiving vessel was previously filled. In this way he was able to collect and study many gases, including several previously unknown. Priestley, however, continued to speak of them as different kinds of air, and he appears to have clung to the idea that the different properties exhibited were due to the greater or less amount of *phlogiston* associated with the gas.

The apparatus used for collecting and operating on gases is thus described in the Introduction to vol. i. of his *Observations:*

"For experiments in which air will bear to be confined by water I first used an oblong trough made of earthenware, as *a* (fig. 1), about eight inches deep, at one end of which I put thin flat stones, *bb*, about an inch, or half an inch, under the water, using more or fewer of them according to the quantity of water in the trough. But I have since found it more convenient to use a larger wooden trough, of the same general shape, eleven inches deep, two feet long, and one and a half wide, with a shelf about an inch lower than the top, instead of the flat stones above mentioned. . . . The several kinds of air I usually keep in cylindrical jars as *cc* (fig. 1), about ten inches long and two and a half wide, being such as I have generally used for electrical batteries, but I have likewise vessels of very different forms and sizes adapted to particular experiments."

The study of fixed air was undertaken by Priestley before 1772, and one outcome of his experiments, coupled with the knowledge that this gas could be absorbed in large quantity by water and that its presence in several natural waters such as those of Spa and Pyrmont gave them their pleasant qualities, was the invention of "soda water." This procured for him the award of the Copley Medal by the Royal Society, and much consequent notoriety. But his next discovery of nitrous air (nitric oxide) was a matter of much greater scientific importance, as it became in the hands of Priestley himself, and later of Henry Cavendish, a valuable reagent in Eudiometry for estimating the goodness of air or, as he expressed it, the extent to which it was *dephlogisticated.*

Priestley's Apparatus for Collecting and Testing Gases.
(Frontispiece of *Observations on Different Kinds of Air*, 1790.)

[*Face p.* ?

Priestley remained at Leeds six years. His reputation as a philosopher brought him an invitation from the Earl of Shelburne (afterwards first Marquis of Lansdowne) to reside with him as companion and nominally librarian at Bowood, near Calne, in Wiltshire.

The stipend of one hundred guineas with a house was little enough for the support of the family, now consisting of two sons beside himself and wife, so that he was naturally tempted by the offer of two hundred and fifty with a house and a retiring allowance for life, beside the advantage of abundant leisure for the pursuit of his researches.

Sir William Petty, second Earl of Shelburne, and first Marquis of Lansdowne (1737-1805) was a nobleman of principles too liberal for the times in which he lived. His policy of conciliation toward the American colonies led to his resignation from the position of one of the Secretaries of State, and from 1768 he lived in retirement. It was during this period that he was joined by Priestley, and for some eight years the arrangement seemed to afford satisfaction to both men. Priestley was then in the prime of life, and pursuing his experiments with great activity he isolated in rapid succession after nitric oxide, nitrous oxide, ammonia, hydrochloric acid gas, sulphur dioxide and silicon fluoride. The second volume of his *Observations* contains the account of his famous discovery of *dephlogisticated air*, or, as it was afterwards called by Lavoisier, *oxygen*. This occurred on August 1, 1774.[1] Though this discovery has been described in every elementary text book of chemistry for more than a century, it is worth noting that the method adopted for the isolation of the gas was of Priestley's own invention. The usual mode of applying heat to glass vessels was by means of a candle flame, but Priestley adopted the plan of concentrating the sun's rays by a large lens, in this case twelve inches in diameter. The statue in Birmingham represents the philosopher holding the lens between his finger and thumb, which is manifestly impossible. He had another of these glasses made sixteen inches in diameter which he took with him to America. The mercuric oxide to be heated was introduced into one of the "short, wide, round-bottomed phials," which he was in the habit of using, "filled with quicksilver and kept inverted in a basin of the same."

[1] The whole account of Priestley's experiments has been reprinted by the Alembic Club, No. 7.

matters. Sunday duty at the New Meeting, classes and lectures, editing the *Theological Repository*, and the numerous theological writings which incessantly issued from his pen left him but little leisure. On arriving in Birmingham he naturally joined the Lunar Society, an informal club or association of men interested in scientific pursuits, similar in constitution and in practice to the club or "Invisible college" out of which rose, a century earlier, the Royal Society. It was a small body consisting of no more than eight or ten members who met at the house of each in turn once a month, on the Monday nearest to the full moon, "in order," says Priestley, "to have the benefit of its light in returning home." Hence the name, Lunar Society. It was founded by Matthew Boulton, the partner of James Watt, the engineer, Erasmus Darwin, author of the *Botanic Garden*, Dr. William Small, a physician, and a few of their friends, and it lasted nearly forty years. The members were accustomed to sit down to dinner at two o'clock, and each in turn discussed some subject in connection with literature, or preferably science, without reference to the religious or political principles of any of the members. Priestley thus found himself an honoured member of a circle which included some of the most famous men of science of the time.

The Lunar Society, according to Mr. H. Carrington Bolton, numbered among its members, at one time or another, Thomas Day (1748-1789), a wealthy, eccentric philanthropist best known as the author of *Sandford and Merton*; Dr. William Withering, F.R.S. (1741-1799), botanist and chemist, after whom the mineral witherite was named; James Watt, F.R.S. (1736-1819), inventor of improvements in the steam engine; John Baskerville (1706-1775), inventor of superior type and a famous publisher, and several of the Galton family. Among visitors to the Society in Priestley's time there were Josiah Wedgwood, the famous potter, Sir Joseph Banks, President of the Royal Society, Sir William Herschel, the most distinguished astronomer of the time, John Smeaton, the builder of Eddystone lighthouse, and others.

It has been the custom among the friends and admirers of Priestley [1] to speak of him "as a peculiarly gentle and inoffensive divine. In point of fact, one has but to glance along the titles

---

[1] These quotations are taken from *The Russells of Birmingham*, by S. H. Jeyes (Geo. Allen & Co. 1911). This book contains an interesting account of the Birmingham riots.

proceedings of the rioters were not stopped for many days, owing apparently to the supineness or timidity of the magistrates. Several other houses were burnt, including the mansion of William Russell, the friend of Priestley. In the sequel several of the ringleaders were executed and partial compensation was paid to the sufferers. Priestley's claim for damages amounted to £3,628 8s. 9d., though it is believed that his actual loss was greater. The court awarded him £2,502 18s.

The great body of the clergy of the Established Church at this period were unhappily so affected by bigotry as to lose no opportunity of insulting and abusing Dissenters of all kinds, and their example served to encourage the persecution which pursued the unfortunate Dr. Priestley in his retreat. Attempts to establish himself and build a new laboratory at Hackney were rendered abortive in consequence of renewed threats of violence, and his position even in relation to the Royal Society became so intolerable that he renounced his Fellowship. Ultimately Priestley decided to leave his native country and settle in America. His friends, the Russells, determined on the same course. Priestley sailed from London on April 8, 1794, and reached New York on June 4. The Russells took ship at Falmouth on August 13th, but having been taken prisoner of war by a French frigate, they did not arrive on American soil till the following year.

Priestley soon found many friends in America and was offered the professorship of chemistry in the University of Philadelphia. This, however, he declined, as well as the charge of a Unitarian congregation in New York, and he ultimately decided to settle at Northumberland, a town in Pennsylvania, and there he continued to write and made preparations for carrying on his experimental work. His friendship was maintained with the Lunarians of Birmingham, and especially with Boulton and Watt, who sent him presents of apparatus; but whatever he may have found out in connection with his experiments on air, or whatever information he may have received concerning the experiments of Cavendish and the work of Lavoisier, led to no modification of his views in theoretical chemistry. His last essay, entitled *The Doctrine of Phlogiston Established and that of the Composition of Water Refuted* was published at Northumberland in 1803. In this work he remarks on the almost universal acceptance of the new theory of combustion, and it must ever seem remarkable that one so full of liberal views in every other direction should have shown so tenacious a conservatism in the interpretation of his own observations and results. Of this condition of mind Priestley seems to have been unconscious, for in the preface to the third volume of his *Observations on Air* he remarks :

"Upon this, as upon other occasions, I can only repeat, that it is not my *opinions*, on which I would be understood to lay any stress. Let the *new facts*, from which I deduce them, be considered as my *discoveries*, and let other persons draw better inferences from them if they can."

His health began to fail in 1801, and on February 6th, 1804, he quietly breathed his last. He was buried in the local cemetery.

Although contemporary with Black in Glasgow and Edinburgh, and with Henry Cavendish, who resided the greater part of his life at Clapham, Priestley does not seem to have held communication with either of these philosophers whose scientific pursuits so nearly resembled his own. He was probably acquainted with Cavendish, as he attended many meetings of the Royal Society, and on one occasion, December 10th, 1773, he dined with the Royal Society Club. He was certainly acquainted with Cavendish's work on air and on the composition of water, as he refers to his results only to express dissent from them.

Though through stress of misfortune Priestley was driven to seek refuge in another land, he never ceased to remember that he was born an Englishman, and his sympathies remained with the country of his birth. He was therefore never naturalised as an American citizen, nor did he take part directly or indirectly in any election. The fearless avowal of his principles, however, brought him many enemies even among the American public. On the other hand, the affection and admiration of his friends was constantly manifested by financial assistance and in other ways down to the end of his life. In 1802 a subscription was raised by his friends in England to enable him to print his *Church History* and his *Notes on all the Books of Scripture.* He was, in fact, one of the most prolific writers on record, and at the end of the *Memoirs*, written partly himself and partly by his son, Joseph Priestley, after his death, a catalogue is given of the books written by him, which contains no fewer than one hundred and eight titles.

THE HONOURABLE HENRY CAVENDISH.

The only known portrait.

[*Face p.* 41].

CHAPTER IV

CAVENDISH

While the tragic events arising out of the Revolution were proceeding in France, the rest of Europe was naturally thrown into a state of ferment. France, having abolished the monarchy and proclaimed a republic, was at war with Spain, Holland, Austria and Great Britain. England was overrun with the unhappy *émigrés* who had escaped from France and from the atrocities perpetrated by the revolutionary committees.

As already pointed out, a state of political disturbance or even violent disorders, provided it does not touch the persons of the individuals, never seems to arrest the continuance of their work by students of nature living at the period, and though Priestley had been obliged to leave his native land, investigation of chemical phenomena, and especially those connected with fire and air, was still carried on in this country by others. The most notable figure in the scientific world at this period was Henry Cavendish. Strange to say, very little is known of the early life of this remarkable man, son though he was of a noble family which could look back over eight centuries to Norman times for its first distinguished members. Something more may yet be learned about his work, though probably nothing can now be added to the scanty records of his life when the remainder of the Cavendish papers belonging to the Duke of Devonshire and now under investigation by representatives of the Royal Society have been fully examined.

The Honourable Henry Cavendish was the elder of two sons of Lord Charles Cavendish, who married Lady Anne Grey, fourth daughter of Henry, Duke of Kent. Lord Charles Cavendish was the third son of the second Duke of Devonshire. Henry was thus a grandson of a duke by both parents, and was closely connected with many of the aristocratic families of Great Britain. He was born on October 10, 1731, at Nice, whither his mother, Lady Anne Cavendish, had gone for the benefit of her health. The Honourable Frederick Cavendish, his only brother, was born in London. Nothing is known concerning the earlier years of the brothers beyond the fact that in 1742 Henry was a pupil at Dr. Newcome's School at Hackney, whither his brother was sent a year or two later. From school he went to Cambridge, where he entered at St. Peter's College, and commenced residence on November 24, 1749. He remained in regular attendance till February 23, 1753, and then, although the full time had been spent in residence required for the degree, he left without graduating. Henry's younger brother Frederick was entered on April 10, 1751, and also left without taking his degree. His cousin, Lord John Cavendish, fourth son of the Duke of Devonshire, entered in February, 1750, and took the degree of M.A. in the year 1753. There is no record of what Henry Cavendish did during the next ten years. He appears to have visited Paris together with his brother Frederick, and in all probability he occupied himself in mathematical and physical studies. From the papers he left, which are for the most part undated, it is impossible to say in what direction these studies led him. His first published researches were, however, chemical, and his paper on *Factitious Airs* was published in the *Transactions of the Royal Society* for 1766. This paper consisted of three parts which were published at the time, and a fourth part which remained after his death ready for the press, but unpublished till it was disinterred by the Rev. W. Vernon Harcourt and communicated to the British Association in 1839. These papers contain the results of Cavendish's experiments on carbonic acid, hydrogen, and other gases already known, but imperfectly characterised by previous researches, and insufficiently distinguished by their properties from one another. Inflammable air had long been known, but Priestley and other chemists up to this time do not seem to have been aware that the inflammable air obtained from metals by dissolution in sulphuric or muriatic acid was not the same thing as the inflammable air obtained from charcoal. In the earliest of the memoirs already mentioned Cavendish for the first time showed that the different kinds of air have different specific gravities, and his experiments were conducted with refinements previously neglected, including the use of materials for drying the gas operated on and corrections of volume for temperature and pressure.

The name of Cavendish in the department of chemical as distinguished from physical research will for ever be associated with the study of the properties of hydrogen and the establishment of the composition of water. Under the title *Experiments on Air* two important papers were published in the *Philosophical Transactions* for 1784 and 1785.[1] The former contains an account of the experiments by which the composition of water was proved experimentally and this important point fully established by Cavendish. The combination of inflammable air with common air had already been tried by Mr. Warltire, a friend of Priestley, who found a slight loss of weight and that when the experiment was repeated in glass vessels there was a deposit of dew. This was explained by Priestley by the hypothesis that "common air deposits its moisture by phlogistication." Here it is permissible again to remark on the confusion and complication of language imposed by the employment of the phlogistic doctrine throughout the papers of Cavendish, equally with those of Priestley, in giving an account of his experimental enquiries.

What Cavendish did finally was to prove beyond dispute that when inflammable air from metals is burnt with about 2½ times its bulk of common air, and the burnt air is made to pass through a long glass tube, the only liquefiable product of the experiment, the dew, in fact, is pure water. In the experiment of which details are given by Cavendish no less than 135 grains, or more than a quarter of an ounce of water, were collected. Other experiments were made in which the two airs were fired by an electric spark in a glass globe, which is represented on the modern lecture-table by the pear-shaped vessel commonly called "Cavendish's Eudiometer." The result may be recalled in his own words as follows: "By the experiments with the globe it appeared that when inflammable and common air are exploded in a proper proportion, almost all the inflammable air and near one-fifth of the common air lose their elasticity, and are condensed into dew. And by this experiment it appears that this dew is plain water, and consequently that almost all the inflammable air and about one-fifth of the common air are turned into pure water."

---

[1] These may be read in the Alembic Club Reprint, No. 3 (Edinburgh. Clay).

But this is not all, for Cavendish then proceeded to examine the result of firing a mixture of inflammable air (hydrogen) with dephlogisticated air (oxygen) and obtained, by collecting the products of successive experiments, about 30 grains of water, which, however, was acid to the taste and when saturated with alkali yielded nearly two grains of nitre. Enquiry into the source of this acid led to further experiments in which Cavendish took the precaution of using oxygen from several sources, and also in different experiments different proportions of hydrogen. He found that when the burnt air "was almost completely dephlogisticated" (that is, a relatively large proportion of hydrogen was used), the condensed water was not at all acid, whereas when the proportion of hydrogen used was less, and therefore the proportion of oxygen greater, the amount of acid produced was greater and this acid "was always of the nitrous kind."

Cavendish agreed with Lavoisier and Scheele that common air is a mixture of two distinct substances and that though they were respectively named dephlogisticated and phlogisticated airs they do not differ only by their degree of phlogistication, and he also arrived at the conclusion that *mercurius calcinatus* (red precipitate), however prepared, is "only quicksilver which has absorbed dephlogisticated air from the atmosphere during its preparation." This is, of course, what Lavoisier taught, but although Cavendish was acquainted with Lavoisier's doctrine he did not accept it, and in the paper from which the preceding extracts have been made, he explained why he still preferred to use the language of the phlogistic theory, and this phraseology he continued to employ in a second paper on air communicated to the Royal Society the following year (1785). The second paper gives an account of experiments by which the composition of nitric acid was discovered. It shows that the diminution of volume brought about in atmospheric air by the action of the electric spark alone is not due to the generation of fixed air as had been previously supposed, but that it "depends upon the conversion of phlogisticated air into nitrous acid."

This paper contains the passage which in recent times has become famous in consequence of its relevance to the experiments made a hundred years later by Lord Rayleigh which resulted in the isolation of argon. Cavendish pointed out that in his day very little was known concerning the properties of phlogisticated air (nitrogen), and it was doubtful whether the whole of the phlogisticated air of the atmosphere was of one kind, or "whether there are not in reality many different substances confounded together by us under the name of phlogisticated air." And then he goes on to describe the experiment by which he settled this point. He had already found that a small quantity of acid was produced when dephlogisticated air and inflammable air were exploded together and that this proceeded from the accidental admixture of a small quantity of phlogisticated air. Then he goes on to relate how he "therefore made an experiment to determine whether the whole of a given portion of the phlogisticated air of the atmosphere could be reduced to nitrous acid, or whether there was not a part of a different nature from the rest which would refuse to undergo that change. The foregoing experiments indeed in some measure decided this point, as much the greatest part of the air let up into the tube lost its elasticity; yet as some remained unabsorbed[1] it did not appear for certain whether that was of the same nature as the rest or not. For this purpose I diminished a similar mixture of dephlogisticated and common air in the same manner as before till it was reduced to a small part of its original bulk. I then, in order to decompound as much as I could of the phlogisticated air which remained in the tube, added some dephlogisticated air to it and continued the spark till no further diminution took place. Having by these means condensed as much as I could of the phlogisticated air, I let up some solution of liver of sulphur to absorb the dephlogisticated air; after which only a small bubble of air remained unabsorbed, which certainly was not more than $\frac{1}{120}$ of the bulk of the phlogisticated air let up into the tube; so that if there is any part of the phlogisticated air of our atmosphere which differs from the rest, and cannot be reduced to nitrous acid, we may safely conclude that it is not more than $\frac{1}{120}$ part of the whole."

It may perhaps be worth while to state in modern language what Cavendish had accomplished. He knew that the atmosphere consisted mainly of two kinds of air, the one now called oxygen which combines with hydrogen to form water and unites with mercury (producing red precipitate) and with other metals. The other constituent of air, since named nitrogen, was now proved to be uniform in constitution and was convertible,

[1] By the alkaline solution used in the experiment.

with the exception of the small residue amounting to $\frac{1}{120}$ of the whole, into an acid which was absorbed by alkaline solutions and converted into nitre. This result was arrived at by sparking common air mixed with successive additions of oxygen in a tube containing caustic potash, the surplus oxygen being withdrawn at the end of the experiment by introducing a little alkaline sulphide.

The reader may also be reminded that at the time Cavendish was making his experiments the only known method of procuring electric sparks was by friction of a glass globe or cylinder by means of a pad pressed against it while the cylinder was rotated by hand.

Cavendish had a turn of mind essentially mathematical, and consequently most of his researches assumed a quantitative character. He not only determined the fact that the air of the atmosphere contains two chief constituents, but he desired also to know in what proportions they were present and to what extent, if any, they varied in air taken from different places. Priestley had discovered that when nitric oxide is mixed with air a contraction follows, and was ignorant of the fact that the amount of contraction varies according to the way in which the gases are brought together. He therefore supposed that the purity of the air, that is, the amount of dephlogisticated air (or oxygen) which it contained, could be ascertained by mixing it with a measured quantity of nitric oxide over water and observing the contraction which ensued. The instrument used was called a *Eudiometer* or measurer of the goodness of the air, or, to use the language of the period, its freedom from phlogiston, which was supposed to be the cause of bad quality. Priestley, by employing nitric oxide, had succeeded only in showing that common air contains about $\frac{1}{5}$ of good air, but he found by applying his process that though air from different localities seemed to vary in quality it was not always the hypothetically bad air which was found by the test to be most deficient. Cavendish undertook the examination of the question, and ultimately came to the very important conclusion that atmospheric air is substantially constant in composition. It is unnecessary to describe Cavendish's modification of Priestley's method of analysis, as it has long since been abandoned by all chemists, and an account of it and all the precautions taken would be tedious. Suffice it to say that he succeeded in using it in such a way as to get uniform results and, as a consequence of the long series of

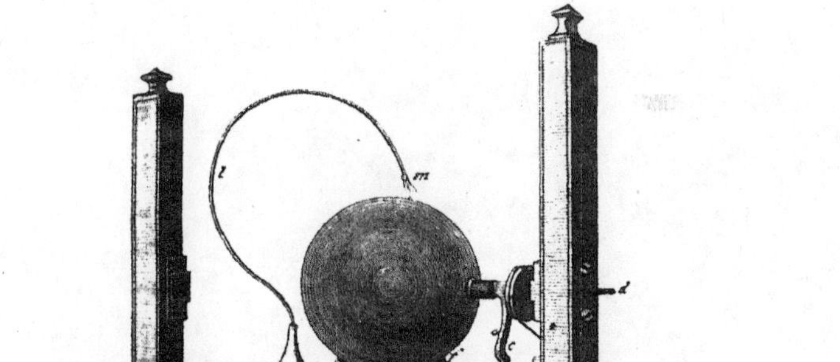

ELECTRICAL MACHINE AS USED BY PRIESTLEY, AND, PROBABLY, BY CAVENDISH.

(Priestley's *History of Electricity*, 1794.)

[*Face p.* 46]

experiments he undertook, the proportions of the constituents expressed in modern language were found to be—

Oxygen	20·833
Nitrogen	79·167
	100·000

And these figures are practically identical with those which result from the best experiments of modern times. The following account is given by Cavendish of his researches on the subject: "During the last half of the year 1781 I tried the air of near sixty different days in order to find whether it was sensibly more phlogisticated at one time than another, but found no difference that I could be sure of, though the wind and weather on those days were very various, some of them being fair and clear, others very wet, and others very foggy." Altogether some hundreds of analyses were made, and as already mentioned the results agree closely with those of modern times, while the experiments of his most eminent contemporaries led to conclusions very different and diverse. Scheele, for example, found the proportion of oxygen 25 per cent., Lavoisier made it 27 per cent. of the air operated on.

Cavendish pursued his chemical researches as part of a general philosophical view of nature, and he made frequent excursions in other directions. Thus in 1771 he published a paper in the *Philosophical Transactions* on the phenomena of electricity, and in 1776 he showed that the effects on the human body produced by the *torpedo*, or cramp-fish of the Mediterranean, are probably produced by electric discharges.

Cavendish's views concerning the theory of quantity and intensity of electricity were held deserving of respect when, more than fifty years later, they were examined by Faraday. Cavendish early became interested in the phenomena of heat, and succeeded in determining with considerable accuracy the melting or freezing point of mercury, concerning which very extravagant ideas had previously been accepted. In 1796-7-8 Cavendish carried out the famous enquiry concerning the mean density of the earth to which his name is usually attached. The apparatus consisted essentially of a long rod suspended by the middle and having at each end a small ball of lead, the whole enclosed in a

distinguished foreigner, who was anxious to make his acquaintance and had come to England with that object in view, Cavendish stood, silent and confounded, till spying an opening in the surrounding crowd—it was at a reception by the President of the Royal Society, Sir Joseph Banks—he rushed out of the room, got to his carriage and drove directly home. Many accounts have been given of his peculiarities. Among the rest memoranda were left by Sir Humphry Davy and Dr. John Davy, his brother, by Professor Playfair, of Edinburgh, by Dr. Thomas Thomson, and others. From these we learn that his appearance was insignificant, and except for the peculiarity of his dress would not have attracted attention. He wore the costume of his grandfathers, the coat of faded velvet with high collar, shirt frilled at the wrist, with three-cornered cocked hat, a fashion from which he never departed. He had a thin squeaking voice, with a hesitation or difficulty of speech which increased when he was embarrassed, as was often the case in the presence of strangers. He encouraged no intimacy with anyone and rarely entertained guests at his house. It was related that Cavendish seldom had company, but on one occasion three or four Fellows of the Royal Society were to dine with him and his housekeeper came to ask what was to be got for dinner. The order was, as usual, "a leg of mutton." "Sir, that will not be enough for five." "Well, then, get two," was the reply.

His father, Lord Charles Cavendish, died in 1783, but even before that time Henry had become a wealthy man. From what source or sources his fortune was derived is uncertain, but at the time of his death he was worth altogether about £1,500,000 and was the largest holder of bank stock in England. Cavendish, however, made little use of his wealth, and was so indifferent to the value of money that the story goes that he quarrelled with his bankers for troubling him by asking what they were to do with the large balance remaining in their hands uninvested. His whole thoughts seem to have been occupied with his philosophical studies and experiments. He was, in fact, "le plus riche de tous les savans, et probablement aussi le plus savant de tous les riches," according to the epigrammatic saying of Biot in the *Biographie Universelle*. In the latter part of his life he had two houses in London, one at the corner of Montague Place and Gower Street, on which a tablet, recently placed on the wall, records the fact, the other in Dean Street, Soho, in

accustomed place every week till little more than a fortnight from his death, which took place on March 10, 1810, in the seventy-ninth year of his age. It is thus obvious that he was but little away from London during the last forty years of his life, and that he must have had remarkably good health to be able to make his appearance so regularly in spite of inclemencies of weather and the complaints which accompany them.

That he was the creature of habit was shown in many little ways, such as always hanging his cocked hat on the same peg. At home his boots were placed against the dining-room door always in one spot, and in one particular position, with the point of his stick standing in one of his boots, and always in the same boot.

At the club Cavendish seems to have overcome his timidity to the extent of inviting guests on many occasions. Some of these may have been candidates for admission to membership, others were Cavendish's own friends, and there can be little doubt that he was on terms of some intimacy with the Rev. John Michell, from whom he received the original model of the torsion apparatus, also with Charles Hatchett, the mineralogist, then (1800-2) a young man.

Notwithstanding his peculiarities, Cavendish was apparently profoundly respected by his contemporaries. Of his intellectual powers there was no doubt, and the importance of his discoveries has been acknowledged not alone by his own countrymen and contemporaries but by the world of science at all times. His life was not necessarily unhappy, in spite of the restrictions he imposed on himself, but his days were spent in measurements of one kind or another, and there is no record of relaxation such as enjoyed by most men from art, literature, music or mere sport. On the other hand, he seems to have been incapable of jealousy or quarrelsomeness, for notwithstanding the dispute concerning the composition and nature of water, about 1784, which at one time excited a great controversy, Cavendish met his great rival, James Watt, in friendly intercourse.

For his brother Frederick he seems to have had a cool kind of affection, and they seldom met. He appears to have been accessible to appeals for help from his ample purse, but it was probably true, as was said of him by a contemporary, "that Cavendish did some good in a very ungracious manner."

He would never sit to an artist, and the only portrait existing is a water-colour sketch of which the original is preserved in the British Museum. This was obtained surreptitiously while at dinner, by arrangement with Sir Joseph Banks, who provided the opportunity to the artist, Alexander, "the excellent draughtsman to the China Embassy."

Concerning his death many different accounts have been given, and they agree only in stating that as he felt the end approaching he desired to be alone. The most credible story is that which was given by Sir Everard Home, the well-known surgeon. The substance of Sir Everard's statement was that Cavendish sent his servant away, "ordering him not to come near him till night, *as he had something particular to engage his thoughts and did not wish to be disturbed by anyone*." The servant, who believed his master to be dying, summoned Sir Everard Home, who hastened to Clapham. Sir Everard stayed with the dying man during the night, and the patient remained tranquil till in the early morning he departed this life. He was buried in All Hallows or All Saints Church, Derby, where Elizabeth Hardwick, an ancestress of Queen Elizabeth's time, had built for herself and her descendants a magnificent tomb. For many generations a kind of public funeral was given to all the Cavendishes, and presumably to Henry, for on the death of his brother Frederick two years later he was buried, according to the *Gentleman's Magazine* for 1812, in the family vault, the corpse being met at the entrance to the town by the Mayor and thirty burgesses in mourning.

CARL WILHELM SCHEELE.
From the Statue in Stockholm.

CHAPTER V

SCHEELE

In the endeavour to recall the conditions under which scientific work was carried on in the eighteenth century the fact must not be forgotten that personal communication between students occupied in the same or closely cognate subjects very often affords a stimulus to further effort and is sometimes a source of new ideas and inspiration. Written communications addressed to societies, associations and academies also have their use, inasmuch as they lead commonly to publication in the form of a printed journal which comes before a wide circle of readers. At this period there were many difficulties in the way of both these methods of communication, and scientific men usually stored up the results of their studies until they had accumulated enough material to make a book. This was to a considerable extent Priestley's plan, as already explained, and as other men commonly resorted to the same method it happened occasionally that a discovery remained unrevealed to the world for some time. Priestley's name is indissolubly connected with the discovery of oxygen in 1774, but it appears certain that the same gas was isolated from several sources, and its chief properties examined nearly two years earlier by a poor apothecary, Carl Wilhelm Scheele, working in Sweden. His *Chemical Treatise on Air and Fire* in which his experiments are recorded was, however, not publshed till 1777, and thus the priority in discovery to which he might have laid claim fell to the lot of another. In a letter to Gahn (the mineralogist) dated February 9, 1777, Scheele states that he had not seen Priestley's book, and complains of the want of literature in Köping, where he then was.

The story of Scheele's life is one of pathetic interest, and many details relating to his experimental work, his struggles against poverty and ill-health and his early death, remained for upwards of a century unknown to the world. The design of collecting together his laboratory note-books, his letters and other papers, was conceived shortly after his death by the secretary of the Swedish Royal Academy of Sciences, Johann Carl Wilcke. This design, however, was frustrated by the death of Wilcke himself in 1796. By the patriotic labours of Baron A. E. Nordenskiöld, assisted by Frau Elin Bergsten, who deciphered and copied the laboratory notes and letters, the work was ultimately accomplished, though nearly a century later, and with the permission of the Swedish Academy these relics were collected into one handsome volume in 1892, with, simultaneously, a translation into German. From this volume full information is obtainable concerning Scheele's life and work.

Carl Wilhelm Scheele was born on December 9, 1742, one of a family of eleven children consisting of six sons and five daughters, Carl Wilhelm being the youngest but one of the sons. His father, Joachim Christian Scheele, was a merchant at Stralsund, the chief town in what was then Swedish Pomerania. Though connected with a highly respectable family, he does not seem to have been in a position to afford his sons an academic education. Two of them adopted the calling of *apotheker*. One of these was Carl Wilhelm, who after a very short period of school life in 1757, at the age of fourteen years entered the business of Martin Anders Bauch at Gothenburg. The elder brother, Johann Martin, had been placed in the same pharmacy, but had died three years before.

Discerning the ability of his young pupil and his inclination toward study, the master afforded liberal opportunities for acquiring knowledge of chemical and pharmaceutical matters. In a small town of ten or eleven thousand inhabitants in those days it may be imagined there were but few opportunities of instruction, but the pharmacy itself afforded material enough for observation and experiment. In course of time every one of the chemical substances with which in his youth Scheele was thus brought into contact became the subject of his observations or the starting-point for discoveries. Among the books to which he had access were Caspar Neumann's *Prælectiones Chemicæ*, Lémery's *Cours de Chimie*, Boerhaave's *Elementa Chemiæ*, von Löwenstern-Kunckel's *Laboratorium Chymicum*, and Rothe's

Anleitung zur Chymie. Such were the aids with which Scheele made himself a competent researcher.

In letters to Scheele's father, his master Bauch always reported of his pupil in terms of the highest praise, and only expressed fear that studies often protracted into the night would injure his health. Similar testimony to the ardour with which he pursued his experiments was given by other persons, including fellow-pupils.

In 1765 Bauch sold his business and Scheele took a situation with the apotheker P. M. Kjellström in Malmö. From this time dates his acquaintance with Anders Johan Retzius, who was about the same age and later became a distinguished professor in the University. The friends corresponded for some time after Retzius left Malmö, and at Scheele's death Retzius published an enthusiastic memorial of his talents and of his extraordinary diligence in the pursuit of experimental investigation. Without systematic training, he worked with no prejudices and untrammelled by theory, so that he observed much and discovered many things which convention would regard as impossible, being contrary to theory. It is not surprising that Retzius should have been astonished at the large number of experiments recorded in his note-books, and science is undoubtedly indebted both to Bauch and Kjellström for the almost unlimited freedom allowed to their assistant in the pursuit of his enquiries.

In 1768 Scheele removed to Stockholm, where, however, he was employed not in the laboratory of the pharmacy but in the shop. His opportunities for experiment were thus restricted, but he made use of a sunny window in which he was able to establish the important fact that the reduction of silver chloride was promoted in different degrees by different portions of the solar spectrum.

Retzius followed Scheele to the capital a few months later, and the result of a research undertaken by the friends jointly was the isolation of tartaric acid from cream of tartar. A paper on the subject appeared in the *Transactions of the Royal Swedish Academy of Sciences* in 1770, and thus for the first time Scheele's name appeared in print as a discoverer. A previous attempt at publication had been unsuccessful, but from Scheele's correspondence with Gahn and laboratory notes which have now been published, it appears that Scheele had by this time found that inflammable air (hydrogen) is evolved when iron or zinc is digested with an organic acid and water, and the laboratory notes contain a drawing of a simple apparatus for collecting the gas.

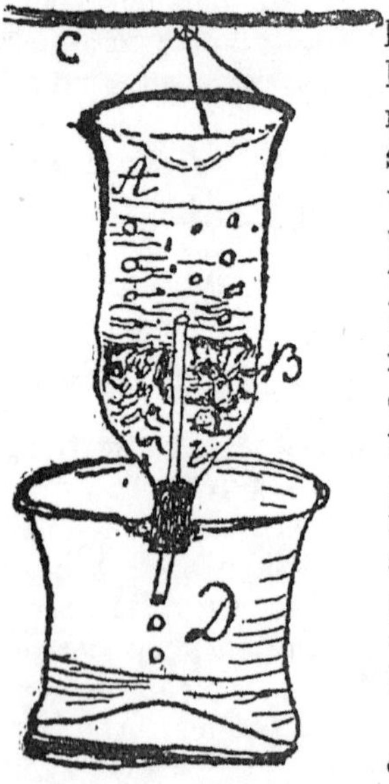

Scheele appears to have thought he had got hold of *phlogiston* itself, which he supposed was a constituent of the metal. He devised an apparatus, shown in the accompanying figure, by which he obtained the gas, not only with the aid of acids, but by means of water and iron filings alone. A is a bottle containing iron filings (B); it was filled with water to the neck, a stopper carrying a glass tube was added and the whole hung up in an inverted position over a glass vessel (D). In a few days bubbles appeared on the iron and in two or three weeks the gas displaced the water. The inflammable air which filled the bottle was pronounced by Scheele to be *phlogiston elasticum.*

Another paper of Scheele's, *Chemical Experiments with Salacetosella* (*potassium acid oxalate*), was also rejected by the Academy. It seems probable that this was due to some omission on the part of Bergman, who at that time was Professor of Chemistry and Pharmacy in the University of Upsala, and to whom the paper was referred by the Academy. This, however, could not have been due to any personal prejudice, as Scheele was probably unknown to him at that time. This acquaintance, however, was accomplished very soon afterwards, for Scheele, in the autumn of 1770, migrated to Upsala, where he took a place as assistant in the laboratory of the apotheker Lokk.

Before leaving Stockholm Scheele had struck up a friendship with Johann Gottlieb Gahn, a mineralogist of about his own age, and with whom he afterwards maintained a correspondence extending over many years. Gahn was the means of introducing the clever young apothecary to the Professor, and it cannot be doubted that this meeting was to the advantage of both. Bergman was a learned and accomplished man, possessing, however, but slight experience in practical chemistry. The duties of the professor in those days did not include much in the form of practical instruction, and the Universities were for the most part destitute of laboratories. On the other hand, Scheele had already distinguished himself by his experimental skill, which he had exercised independently of the theory and the mysterious language still employed in the chemical writings of the day. Retzius was, in fact, led to declare in reference to the relations between the two men, that it was difficult to decide which was the teacher and which the taught. Bergman continued to write on a variety of subjects till his death in 1784, but though he undoubtedly rendered good service to science in laying the foundation of quantitative chemical analysis, and in the application of chemistry to mineralogy, it will be doing no injustice to his memory to repeat the saying of one of his biographers, that "the greatest of Bergman's discoveries was the discovery of Scheele."

Scheele, however, learned much from the Professor, and it appears that it was at Bergman's suggestion that he undertook the investigation of *black magnesia* (or manganese), which led to such epoch-making results. Fortunately Scheele seems to have had abundance of time for research while in Lokk's laboratory, and while in Upsala he completed his great work on *Fire and Air*. Here before 1773 he had isolated oxygen by the action of heat on silver carbonate, mercuric carbonate, mercuric oxide, nitre, magnesium nitrate and by distilling a mixture of arsenic acid and manganese dioxide. It is probable even that long before in Malmö he had obtained the gas in connection with his examination of nitrous acid: the so-called *salpeterluft*, the red oxide of nitrogen, and its properties were known to Scheele long before anything had been published on the subject. The examination of manganese led not only to the isolation of oxygen, but of chlorine, arsenic and baryta, and the recognition of manganese as a metal previously unknown. The chemical characters of silica, magnesia, microcosmic salt and oxalic acid were also demonstrated by Scheele.

Hitherto academic professional prejudice had prevented Scheele, a mere student of pharmacy, from taking part in University life, but on February 4th, 1775, at a meeting at which the king was present, Scheele was elected a member of the Royal

Academy of Sciences, an honour which never before or since had been conferred on a student of pharmacy.

In order now to exchange his subordinate position in Upsala for one of more independent character, Scheele accepted in 1775 the appointment of superintendent of the pharmacy of Köping, a small town some sixty miles from Upsala, in succession to Hinrich Pascher Pohls, whose privileges on his death passed to his young widow, Sara Margaretha Sonneman. The change was very welcome to Scheele, who in a letter to Gahn burst forth, "Oh, how happy I am! No care for eating or drinking or dwelling, no care for my pharmaceutical business, for this is mere play to me. But to watch new phenomena this is all my care, and how glad is the enquirer when discovery rewards his diligence; then his heart rejoices."[1] Unhappily it turned out that the peaceful life he had looked forward to was not as yet to be his. The widow and her father entered into negotiations for the sale or lease of the business to another man. This appears to have affected Scheele deeply and, little as he was accustomed to discuss his private affairs, his letters at this time reveal his disappointment.

However, Scheele received from many quarters evidence of the high esteem in which he was held. Gahn invited him to join him at Falun. Bergman desired him to return to Upsala. It was proposed that he should take charge of the pharmacy at Alingsäs with a private laboratory attached, and it was publicly suggested that he should return to Stockholm with the title of *Chemicus regius*. He even received offers from foreign countries. D'Alembert, in a letter of December 15, 1775, addressed to the King of Prussia, proposed that he should be called to Berlin, and according to a letter from his brother Christian Scheele to Wilcke, preserved in the library of the Royal Academy of Sciences, he actually received in 1776 an invitation with an offer of a salary of 1,200 reichs-thalers. It is also stated in Crell's

[1] Letter to Gahn, December 26, 1774: "Hochgeehrter Herr, mein werthester Freund! O wie glücklich bin ich! Keine Sorge für Essen und Trinken, keine Sorge für Wohnung, keine Sorge für meine pharmaceutischen Laborationen und für die Anfertigung der Defecta—denn dies ist nur ein Spiel für mich. Aber neue Phenomena zu erklären, dieses macht meine Sorge aus, und wie froh ist der Forscher, wenn er dass so fleissig Gesuchte findet, eine Ergötzung wobei das Herz lacht."

Biographie that overtures from England had placed at his disposal £300 a year, with an easy and advantageous situation. This statement, however, was altogether discredited by Thomas Thomson in his well-known *History of Chemistry*. After enquiry of Sir Joseph Banks, Mr. Cavendish and other chemists in Great Britain, none of them were found to have ever heard of such negotiation. And that such a proposal should ever have been made is improbable. As remarked by Thomson, "while in every other nation in Europe science is directly promoted and considerable sums are appropriated for its cultivation and for the support of a certain number of individuals who have shown themselves capable of extending its boundaries, not a single farthing has been devoted to any such purpose in Great Britain. Science has been left entirely to itself; and whatever has been done by way of promoting it has been performed by the unaided exertions of private individuals." Unhappily this was very true, not only in the eighteenth century under the government of George III., but down to times much more recent and almost within living memory.

None of these various friendly proposals was accepted by Scheele, and encouraged by the high consideration he had won for himself among the inhabitants, he preferred to remain in Köping. As he wrote to Bergman: "I cannot eat more than enough, and so long as I can obtain enough I need not seek my bread elsewhere." His neighbours, however, headed by the chief man of the province, procured for him the privilege of opening an independent pharmacy, considering that in allowing him to leave Köping a reproach would be cast alike against the honour and advantage of the town, and they declared their determination to deal with no other apotheker than Scheele.

The sale of the business was thereupon put a stop to and Scheele remained in Köping.

The *Chemical Treatise on Air and Fire* was the great work of Scheele's life. This work was based on the results of experiments carried out partly before 1770 in Malmö and Stockholm, and partly during the first part of his residence in Upsala, that is, before 1773. The manuscript was in the hands of the printer toward the end of 1775, but owing to his dilatoriness it was August, 1777, before the first copies of the book could be distributed to Scheele's friends.

In October, 1776, then, Scheele was finally established in

command of the pharmacy, though this had not been accomplished without a struggle, for he found the business encumbered with debt. This he succeeded in gradually paying off, and a few years later was able to build for himself a new house and laboratory. The widow of Pohls, his predecessor, assisted by one of his sisters, brought over from Stralsund, managed the dwelling-house, and here he continued to live peacefully and happily during the few remaining years of his life. He seems to have made no important journey from Köping after 1777, not even to attend a meeting of the Academy in Stockholm. He was, however, glad to receive visits from friends, especially those engaged in scientific work. As to himself, he devoted his whole life to the laboratory, where his reckless ardour constantly exposed him to danger of various kinds. It is surprising that he did not meet with death in some of his dealings with such poisonous gases and vapours as prussic acid and other cyanogen compounds, and it seems certain that his naturally good health was gradually undermined by frequent exposure to an atmosphere charged with such compounds. Up to the age of thirty-five Scheele had enjoyed good health, but he then began to suffer attacks of rheumatism, probably due to exposure to cold in the shed which served as laboratory during the earlier months of his life in Köping. In the autumn of 1785, however, he was attacked with a more persistent disorder of a similar kind, which in a letter he ascribed to gout, "the trouble of all apothecaries." Notwithstanding the pain he suffered, and the frequent fits of despondency which he found even more hard to bear, he still worked on, and in February, 1786, he sent to the Academy a memoir on gallic acid. In March he was recording his observations on the decomposition of nitric acid in sunlight: "I shall repeat these experiments in the summer, and then we shall see how they turn out." But he never saw that summer. His illness, accelerated by a complication of disorders, ended his life on May 26, 1786. The work of the great investigator was thus brought to a close before he had reached his forty-fourth year. Probably he felt his end approaching, for in his anxiety to provide for the widow Pohls who had performed for him to the last the simple duties of housekeeper, his sister having died in 1780, he not only prepared a will in which he left her sole heir to all his little possessions, but two days before he closed his eyes he made her his wife. The inheritance included the apothecary's privilege, which two years later she transferred to a provisor, Mathias Bölckon, whom she took as her third husband.

It is not difficult to estimate the importance of Scheele's work. He lived at a time when the study of chemical changes and the isolation of elements and compounds and their recognition as distinct substances with characteristic properties were much needed. Without such knowledge chemical theory could make no progress, and the fact that Scheele continued to use the language of the phlogistic doctrine was of comparatively small importance in view of the host of definite discoveries which he established and which thus placed material for further work at the disposal of the chemist. Not only did he isolate oyxgen and chlorine, beside other substances already mentioned, including many of the vegetable acids such as tartaric, citric, lactic, benzoic, gallic, oxalic and others, but he showed the general character of the oils and fats by discovering what he called the sweet principle of oils or glycerin by saponifying lard with oxide of lead. He also isolated the sugar of milk. He was the first to show the nature of fluorides by preparing the corresponding acid, though in an impure state, and he discovered prussic acid and the means of producing Prussian Blue. The retention of the name Scheele's Green is an acknowledgment of his work on compounds of arsenic, as the survival in pharmacy of the designation Scheele's Acid applied to the stronger solution used occasionally in medicine down to comparatively recent times is testimony to his work in that direction. He learned to recognise tungstic, molybdic and arsenic acids, and he demonstrated the true nature of plumbago as nothing more than impure carbon.

While through these years all this work and these discoveries of Scheele were in progress it must not be forgotten that, though in distant lands, there were other experimenters busy at the same time on kindred problems. In England we have seen that Black and Priestley in pursuit of their enquiries had been led to important discoveries.

In the meantime the volume of Scheele's letters includes one from the Swedish chemist to his illustrious contemporary Lavoisier in Paris, which serves to remind the reader how isolated were these solitary workers in those days and how few were the opportunities of comparing their results or of obtaining assistance from one another. On this occasion Scheele offers his thanks to Lavoisier for a copy of his *Opuscules Physiques et Chimiques*,

(Paris, 1774), and begs him to try an experiment which consists in heating dry silver carbonate by a burning glass and leaving the evolved gas in contact with water, holding lime in suspension. "You will see," he writes, "how much air is produced in which a candle will burn and an animal will live." At this time Scheele was familiar with the properties of oxygen.

Portraits de M^r. & Mme. Lavoisier

d'après le tableau de David

[Face p. 63]

Group III

THE ANTIPHLOGISTIC REVOLUTION

Lavoisier (1743-1794)

CHAPTER VI

LAVOISIER [1]

To appreciate even imperfectly the influence of Lavoisier on the progress of science it is necessary to recall to mind some of the circumstances, political and social, in the midst of which his fifty-one years of life were passed. At the time of his birth, August 26, 1743, Louis XV. was on the throne of France, and in 1774 was succeeded by his grandson Louis XVI. Nearly all the countries of Europe had long been ravaged by war, and famine lurked everywhere, while taxation became more and more oppressive even in England, and especially in France, where the misery of the peasantry has been repeatedly described by historians and made the subject of romance. The great French Revolution, as is well known, was not due to a sudden outburst of unreasoning fury, for it had been preceded during many years by riots and insurrections of the peasants. It has also to be remembered that the world had been prepared for a new order of things by the writings and influence of such men as Rousseau and Voltaire and by the circumstances of the revolution in the American colonies. Nor was the wretchedness of the mass of the people a condition to be found only in France. Here in England the use of wheaten bread by the poor was generally the exception, while mob violence was a frequent testimony of the discontent as well as ignorance and brutality of the majority of the populace alike in town and in country.

Such a state of social disorder seems, however, to have been not inconsistent with considerable intellectual activity among the middle and upper classes of each country, and though art, except perhaps in the direction of music, made but little progress,

[1] The following account is adapted from a lecture given to the Oxford Summer Meeting, August, 1913, by the author.

literature, philosophy and science can each claim to have made substantial growth in the eighteenth century. Almost all knowledge then existing of nature, except as regards the movements of the heavenly bodies, was merely qualitative. To weigh or to measure was an operation as yet rarely resorted to in connection with chemical experiments or in observations concerning the effects of light or heat. The mariner's compass had long been known, but electricity was known only in the form of sparks which could be drawn from various rubbed bodies as well as from the clouds, as shown by Benjamin Franklin in 1749. What we call an electric current was then unknown, and was only discovered half a century later.

The middle of the eighteenth century saw a new and great activity in the investigation of the nature and properties of the atmosphere in which all things at the surface of the earth are immersed. The pressure of this atmosphere had been demonstrated a hundred years previously by Torricelli, and Robert Boyle's experiments on what he called the "spring of the air" had given the law known by his name. But, while some of the mechanical properties of the air were known, its chemical composition and its participation in the processes of burning, of rusting, of decay, and fermentation were practically unknown.

Stahl's famous doctrine of *Phlogiston* was still almost universally accepted, and it remained in possession of the field till driven out by the new preaching of Lavoisier toward the end of this period.

Conditions favourable to the cultivation of scientific pursuits were very different from those which prevail at the present day. Special societies, such as the Chemical, Physical, and Geological Societies did not come into existence till much later, and though the Royal Society in London and the French Académie Royale des Sciences had been long founded there was but little direct encouragement given to scientific work. Moreover, the means of communication between different countries were very imperfect, and though there was doubtless some personal correspondence between men interested in the same pursuits, the publication of the results of scientific research by the Academies of Europe was often much delayed.

The family Lavoisier came from Villers-Cotterets, and though of humble origin, gradually rose in the social scale during the seventeenth century. The father of the chemist, Jean Antoine, was a lawyer, but though without fortune he had married a daughter of the wealthy family Punctis. Young Lavoisier was accordingly educated at the Collège Mazarin under the most favourable circumstances. After studying law for a time he found out the attractions of mathematics and the physical sciences, and at the age of twenty he determined on the pursuit of Science. It will be sufficient to say here that his early efforts were rewarded by his admission to the Academy of Sciences in 1768. About the same time a step was taken which was destined twenty-six years later to lead him to the scaffold.

He became a member of the "Ferme Générale" in the capacity of assistant to the *fermier général* Baudon, who gave up to him one-third of his interest in the lease. This brought him a large income in addition to the moderate fortune inherited from his mother. Lavoisier appears to have regarded this as a means to the end which was the prime object of his life—the advancement of science. He certainly spared no expense in the equipment of his laboratory and in the pursuit of his experimental researches.

The Ferme Générale was a financial company empowered to collect all kinds of taxes, such as the duty on tobacco, salt and imported goods, on condition of paying to the state an annual sum agreed upon in advance for a term of years. Such a system was obviously open to great abuses. The agents of the Ferme often enriched themselves by oppressive exactions, and some of the *fermier généraux* set public opinion at defiance in the scandalous extravagance of their luxury.

It is not surprising that the Ferme was hated throughout France. In Lavoisier's time, under the administration of Turgot, some reforms were introduced, and there is good reason to believe that Lavoisier was among those who discharged their public functions honestly and with the desire to ameliorate as far as possible the hardships imposed on the people from whom the duties had to be collected.

In 1771 Lavoisier married Marie Anne Pierrette, only daughter of the Fermier Général Jacques Paulze. Lavoisier was then twenty-eight years of age, while his *fiancée* was fourteen. Though so young she had great force of character and worked incessantly to improve her knowledge and qualify herself to assist her husband in his researches. She learnt English and translated several works on chemical subjects. She was also a skilful artist and engraver. She made the plates for Lavoisier's *Traité de Chimie* published in 1789, and studied painting under the direction of David.

Arthur Young, the famous English agriculturist, in his travels in France visited Lavoisier, and on October 16, 1787, made the following entry in his diary: "Madame Lav., a lively, sensible, scientific lady, had prepared a *déjeuner anglais* of tea and coffee, but her conversation on Mr. Kirwan's essay on Phlogiston, which she is translating from the English, and on other subjects, which a woman of understanding, that works with her husband in the laboratory, knows how to adorn, was the best repast."

Eleven years after the death of her husband Madame Lavoisier married (in 1805) Count Rumford. She died in February, 1836, at the age of seventy-eight years.

Two characteristics of Lavoisier's work appear in all his published memoirs. He seems always to have in view some great and far-reaching principle, and is rarely interested in detached facts. The greater part of his experiments are quantitative in method; that is to say, he is never tired of weighing and measuring. This is true of even his first communication to the Academy of Sciences in 1765, in which he describes his experiments on gypsum and explains the setting of plaster of Paris when mixed with water. But it is still more obvious in all those numerous researches which occupied many years, in which he investigated the nature and properties of air and water and was led to the complete establishment of those views of his which form the foundation of a system of chemistry which lasted for more than half a century.

In 1774 he published a volume under the title *Opuscules Physiques et Chimiques*, the first part of which is devoted to an examination of the facts which had then been established relatively to the properties and effects of various kinds of air. And the first place is accorded to the work of Black on *Fixed Air*, in which the Edinburgh Professor demonstrated fully and conclusively the cause of the difference between chalk and quicklime, between common and calcined magnesia and between caustic alkali and mild alkali. In all these cases Black had shown that the caustic character of the substance was removed by combination with the gas which we now call carbon dioxide, or carbonic acid, and that this kind of air could be transferred from mild alkali to quicklime or from magnesia alba to quicklime, and that it was set free again either by the action of heat or of acids. That this discovery was considered to be of great importance by Lavoisier is shown by the expressions he used when sending to Dr. Black a copy of his *Opuscules*. He says, in writing to Edinburgh: "Vous avez le bonheur de posséder parmi vous le sçavant Mons. Black qui le premier a réuni en un corps de doctrine le phénomène de la fixation de l'air dans les corps." And it is even probable that Lavoisier's subsequent discoveries were due to the meditations inspired by the knowledge of Black's results, for fifteen years later, sending to him a copy of his *Traité de Chimie* he writes: "Vous y trouverez une partie des idées dont vous avez jeté les premiers germes." The volume of *Opuscules* also contains a historical account of the fact established almost a century previously by Rey, Robert Boyle and others, that metals when calcined increase in weight. It also describes Priestley's work published in 1771 on different kinds of air. These included fixed air produced by fermentation or effervescence, common air in which candles or sulphur had been made to burn, inflammable air produced by dissolving metals, especially iron and zinc, in vitriolic acid diluted with water, or by heating vegetable substances or coal in an iron pipe. Here, of course, Priestley confused together several distinct substances which agreed only in the one property of inflammability.

The second part of Lavoisier's *Opuscules* contains an account of his own experiments, and is entitled *Nouvelles Recherches sur l'Existence d'un fluide Élastique fixé dans quelques Substances, et sur les phénomènes qui résultent de son dégagement ou de sa fixation*. This volume represents the only accomplished portion of a project which the author had formed in 1773. It was not followed by any further volumes of the same character, and it was not till 1789 that Lavoisier brought out his famous *Traité Elementaire de Chimie* in which his new theory was set forth in systematic order.

One of the earliest investigations undertaken by Lavoisier related to the current belief in the transformation of water into earth. And his procedure in this case may serve as a good example of the method employed in many of his researches. Placing water in a glass vessel, called a *pelican* from its general resemblance to the bird of that name, and applying heat to the lower part, the vapour of water rising into the upper part of the neck was condensed and flowed back into the body of the vessel.

Heat was thus applied night and day for more than three months. At the end of that time the weight of the pelican, having been ascertained carefully before the experiment, was found to have diminished. On the other hand the water being evaporated to dryness left an earthy residue the weight of which corresponded exactly to the loss of weight from the glass. Consequently, concluded Lavoisier, water is not changed into earth, but by prolonged boiling it attacks the glass and dissolves a little of it. By such a simple process was the illusion which previously beset chemists at once dispersed. The principle of the method was the judicious use of the balance, and possibly had this instrument found more frequent use among his predecessors some of the discoveries which were reserved for Lavoisier would have been earlier revealed to the world.

Of his contemporaries Black alone systematically made use of the balance, and the important facts he established sufficiently testify to the success which in his hands attends the careful observation of weight in studying chemical change.

At this time it must be remembered that practically everyone employed the theory of phlogiston, with little variation except the temporary hypothesis as to a supposed principle of causticity called *acidum pingue.* The question soon arose in Lavoisier's mind whether the facts observed by himself and others in regard to the reduction of metallic calces by charcoal could be explained by the doctrine of phlogiston. As he says at the end of one of his chapters, time and experience alone will serve to settle opinions on this question. But before reaching the end of the volume it is obvious that Lavoisier had established facts enough to justify the rejection of the idea that when a metal is heated and converted into its calx anything material is lost in the operation. On the contrary, he showed that not only do these *substances increase in weight* in the process, but that the *air diminishes* in amount when the experiment is made in a jar or other closed vessel, and that *this diminution is practically proportional to the increase in weight of the metal.*

Further he showed that there is a limit to the capacity of the air in allowing the calcination of a metal heated in it to proceed, and he concludes that the air we breathe consists of two ingredients, one of which combines with the calcined metal and the other does not so combine.

In a note to this chapter he states that Priestley appeared

not to have suspected that the process of calcination was attended by an absorption or fixation of the elastic fluid.

In 1772 Lavoisier deposited with the Academy a sealed note in which he announced the discovery that sulphur and phosphorus in burning increase in weight, and his conjecture that the increase of weight observed in calcining metals is due to the same cause. He also showed later that phosphorus burned in a limited supply of air uses up only a portion of such air and that the remainder is incapable of allowing phosphorus to burn in it, though strongly heated. Also that this residual air, which consists mainly of what he afterwards called *azote*, extinguishes the flame of a candle like fixed air but that it is lighter than fixed air.

Such was the position when, on August 1, 1774, Priestley obtained oxygen for the first time by his famous experiment in which red precipitate of mercury was heated by the rays of the sun concentrated by a large glass lens. In October of this year he was in Paris and gave to Lavoisier and his friends an account of the remarkable property of this new air in sustaining the flame of a candle, to which air in accordance with his phlogistic ideas he gave the name "dephlogisticated air," that is, air deprived of phlogiston. A month later this experiment was repeated by Lavoisier, and together with a complete account of the properties of the gas was published in the *Mémoires* of the Academy of Sciences for 1775 under the following significant title, *Mémoire sur la Nature du Principe qui se combine avec les metaux pendant leur calcination et qui en augmente le poids*. It is not necessary to revive the now extinct controversy as to whether the discovery of oxygen was made independently by Lavoisier, because the most eminent French chemists have long since agreed that the isolation of this element was undoubtedly first accomplished by Priestley. Indeed so much seems to have been admitted by Lavoisier himself, as in a passage of his Memoir on Nitric Acid he says: "Une partie des experiences contenues dans ce memoire ne m'appartiennent point en propre; peut-être même, rigoureusement parlant n'en est il aucune dont M. Priestley ne puisse reclamer la première idée: mais comme les mêmes faits nous ont conduits à des conséquences diamétralement opposées, j'espère que, si on me reproche d'avoir emprunté des preuves des ouvrages de ce célébre physicien, on ne me contestera pas au moins la propriété des conséquences."

But after a careful study of the papers written by Lavoisier himself it will be clear to the reader that he was already on the track of this important discovery when he met Priestley, and whether he received from the English philosopher any useful information or not he immediately perceived the significance of the discovery; and the considerations which led him to heat *mercurius calcinatus* (red oxide of mercury) by itself (and which are set forth in the memoir referred to above) are more creditable to his judgment and perspicacity than are those which influenced Priestley. For the latter declares that at the commencement of his experiments he "was so far from having formed any hypothesis that led to the discoveries he made in pursuing them that they would have appeared very improbable to him had he been told of them."

Lavoisier, on the other hand, seems to have been led forward step by step, to examine not only the products of the action of heat on various substances in the presence of air, but to have studied the properties of the residual air and to have connected by quantitative measurements the one with the other. Thus, as he says, the metallic calces are reduced to the state of metal for the most part only when heated in contact with some carbonaceous matter or with some substance, which contains what was called phlogiston. And before heating the *marcurius calcinatus* alone he assured himself that when heated with charcoal it behaved like other metallic calces and gave off the same kind of fixed air as the rest. As it was already known to be volatilisable when heated it now only remained to find out whether, when heated without addition of any other substance, it gave forth any kind of elastic fluid, and to find out the properties of the same. This is what Lavoisier did in November, 1774, and repeated with various precautions in the laboratory of his friend Trudaine at Montigny.

With regard to the constitution of water, all the world attributes the discovery of the facts to Cavendish. But the credit of the explanation is ascribed also universally, and justly, to Lavoisier.

Combustion, then, was explained by Lavoisier as a process of combination with oxygen; the formation of the calces of metals and the production of acids were equally attributed to the entry of oxygen into chemical combination with a metal or a non-metal, and similarly the processes of acetification, the nature of the vegetable acids and the changes in the process of respiration were all explained by the action of this principle drawn from atmospheric air. The doctrine of Phlogiston was no more.

Reference has been made to the sealed note deposited with the Academy in 1772. Twenty years later, after the long series of researches in which he finally established his own antiphlogistic views, Lavoisier formed the project of making a collection of his memoirs for publication. But this project was never carried into effect, for the publication was interrupted by his arrest and condemnation; but among his papers was found and printed after his death a sort of introductory essay in which the history of the questions which had so long been debated is set forth. The concluding paragraphs of this essay deserve to be quoted in full, as giving not only Lavoisier's own opinion of his claim to recognition but as showing concisely and clearly what has long been admitted by the scientific world. He says:

"En rapprochant cette première notice de celle que j'avais deposée à l'Académie, à sa séance publique de Pâques 1773, enfin de ceux que j'ai successivement publiés, il est aisé de voir que j'avais conçu, dès 1772, tout l'ensemble du système que j'ai publié depuis sur la combustion.

"Cette théorie, à laquelle j'ai donné de nombreux développements, en 1777, et que j'ai portée, presque dès cette époque, à l'etat ou elle est aujourdhui, n'a commencée à être enseignée par Fourcroy que dans l'hiver de 1786 à 1787; elle n'a été adoptée par Guyton-Morveau qu'à une èpoque posterieure; enfin, en 1785 Berthollet écrivait encore dans le système du phlogistique. Cette théorie n'est donc pas, comme je l'entends dire, la théorie des chimistes français, elle est *la mienne* et c'est une propriété que je réclame auprès de mes contemporains et de la posterité. D'autres, sans doute, y ont ajouté de nouveaux degrès de perfection, mais on ne pourra pas me contester, j'espère, toute la theorie de l'oxydation et de la combustion; l'analyse et la decomposition de l'air par les métaux, et les corps combustibiles; la théorie de l'acidification; des connaissances plus exactes sur un grand nombre d'acides, notamment des acides végétaux; les premières idées de la composition des substance végétales et animales; la théorie de la respiration à laquelle Séguin a concouru avec moi. Ce recueil presentera toutes les pièces sur lesquelles je me fonde, avec leur date; le lecteur jugera."

In the meantime, however, Lavoisier produced in 1789 his famous *Traité Elem. de Chimie*, and the exposition of principles there set forth is so clear, so logical and, with a few trifling exceptions, so fully verified by the work of later times as to leave no room for any sentiment but those of admiration for the genius of the author.

The substitution of Lavoisier's theory of chemical action for the phlogistic theory of Stahl naturally led to a complete revision of the language of chemistry, and the conception of a new system of nomenclature. The name oxygen, *la principe oxigine*, to give it the form originally used by Lavoisier, has long been familiar, but we owe to Lavoisier, associated with de Morveau, with Berthollet and with Fourcroy, all of whom became ultimately convinced of the truth of his theories, the system of names which has almost without modification been in use ever since. Thus they taught the use of the words oxide, sulphide, phosphide for the compounds of the elements oxygen, sulphur, phosphorus. The names of elements are designed to recall the most characteristic properties of each substance without involving the adoption of any theory, and so we get oxygen the producer of acids, hydrogen the producer of water, azote the gas which forms the irrespirable portion of air. Among acids we have changes of termination to indicate the proportions of the acidifying principle, as, for example, sulphuric acid and sulphurous acid, nitric acid and nitrous acid, etc. Similarly the salts of these acids were named sulphates and sulphites, nitrates and nitrites, and so forth.

To Lavoisier also we owe the invention of the word "caloric" for the agent which is the cause of heat, then and long afterwards supposed to be a subtle fluid substance.

A very important line of research was that undertaken by Lavoisier in association with Laplace, the famous mathematician and physicist, in the measurement of the heat given out in various chemical changes by means of the ice calorimeter invented by Laplace. Another series of researches undertaken by Lavoisier with the assistance of Séguin, then a young man, demonstrated the production of animal heat in respiration and the evolution of carbon dioxide. The arrangements in the laboratory are shown in two drawings made by Madame Lavoisier.

One other important departure along a new road is

Lavoisier's Experiments on Respiration. Madame Lavoisier at the Table.

[Face p. 73]

represented by Lavoisier's successful attempts to ascertain the composition of organic substances. This he did by applying the facts already discovered as to the products of the burning of carbon and hydrogen in air; and recognising that in the combustion of spirit of wine, of oils, of sugar and other substances in air, carbon dioxide and water are formed, he arranged an apparatus in which weighed quantities of such substances could be burned and the products collected and weighed, and from their weight the quantities of carbon and hydrogen in the original substance calculated. This principle is the basis of the modern process of combustion introduced by Liebig fifty years after Lavoisier's time.

Lavoisier's services to the State deserve more than a passing mention, but for details the reader must be referred to the work of Grimaux (*Lavoisier*, 1743-1794. Paris, 1888). His work in connection with the Ferme has already been referred to, but in 1775 the *Régie des Poudres* gave him another set of duties which absorbed much time and attention. In 1785 he became a member and soon afterwards secretary of the Committee of Agriculture, and in 1787 he was chosen as a member of the *Assemblée de l'Orléanais*, a sort of County Council empowered to regulate the local taxes. This gave Lavoisier an opportunity of attempting, though unsuccessfully, to relieve the peasantry of the unjust and oppressive *corvée* which compelled the inhabitants of the country to make and maintain the roads and to supply the means of transport in horses and vehicles.

In 1791 a commission was appointed for the purpose of contriving a new and uniform system of weights and measures to supersede the scandalous diversity which prevailed not only in every province but in almost every parish in the kingdom. The metric system was accordingly devised, and the National Assembly having voted a considerable sum of money to defray expenses, Lavoisier was appointed treasurer. The preparation of the necessary instruments occupied some time, and work was not commenced till 1792. In the meantime the fate of the Academy of Sciences, of which Lavoisier was also treasurer, was being gradually determined by the forces operating as the result of the Revolution. Meetings continued to be held, but though Lavoisier took an active part in all the proceedings, his own researches in the laboratory were necessarily interrupted. During the year 1793, however, the academies and learned societies, legacies of the monarchical régime, were crumbling to pieces one after another.

The king had suffered death on January 21, 1793. The payments to the academicians, to which many of them looked as their sole means of support, were interrupted, and on August 8th the Convention decreed the suppression of all learned societies, including the Academy. This catastrophe was hastened on by the manœuvres in the Convention of the chemist Fourcroy. The son of a *pharmacien*, he had been since 1784 professor of chemistry at the Jardin des Plantes, and in 1785 was elected a member of the Academy of Sciences, from which he had applied for assistance in some of his researches. Apparently disappointment and perhaps jealousy of Lavoisier animated this man with a determination to effect the destruction of the Academy. In September a new commission of weights and measures was appointed to supersede the control by the Academy. Lavoisier throughout all these dark days had exerted himself by every means in defence of the Academy and the interests of Science. But now the clouds were descending on his own head and he found himself among those suspected of "incivism" and subject to domiciliary visits. Nevertheless his efforts did not relax, and the two months of liberty still remaining to him were devoted to carrying on the great work begun by the Academy. Nor was he occupied only with the scientific work of the commission of weights and measures. The organisation of public instruction had been entrusted to him by the *Bureau de Consultation des Arts et Métiers,* established in 1791, and the whole scheme was framed by Lavoisier, who took part in all the meetings of the bureau up to the day of his arrest, on November 28, 1793, the very day on which he was elected President.

Meantime the enemies of Lavoisier were at work. Marat, at the commencement of his career, had hoped to make a name for himself in connection with physical science, and could not forget the contempt which had been poured on his pretensions to a theory of fire in 1780. Accordingly he could not now refrain from displaying his animosity in violent denunciations of the most eminent of the academicians. Lavoisier was, however, the mark against which his vengeance was specially concentrated. Fourcroy, the chemist, himself a former academician, having already succeeded in overthrowing the Academy, now did not hesitate to attack its most illustrious member as a counter revolutionary. It was this same Fourcroy who pronounced the *éloge* of Lavoisier in the course of the theatrical obsequies intended to do honour to his memory the year after his death. In the history of science a more despicable character is scarcely to be found.

The condemnation of the *fermiers généraux* was now hurried on. They were lodged in the Port-libre prison and were required immediately to render account of the financial operations of the Ferme. Lavoisier and his father-in-law Paulze were among the prisoners. The former occupied himself with preparing his scientific papers for the press, while his wife exerted herself in every direction in search of help. Lavoisier begged her to take care of her health and not to exhaust herself in useless efforts. The only privilege accorded her was to visit her husband in the prison. A letter written to his wife on December 19, 1793, shows the anxiety only half expressed which filled his mind:

"Tu te donnes, ma bonne amie, bien de la peine, bien de la fatigue de corps et d'esprit, et moi je ne puis la partager. Prends garde que ta santé ne s'altère, ce seroit le plus grand des malheurs. Ma carrière est avancée, j'y ai joui d'une existence heureuse depuis que je me connois, tu y as contribué et tu y contribues tous les jours par les marques d'attachement que tu me donnes ; enfin je laisserai toujours après moi des souvenirs d'estime et de consideration. Ainsy ma tâche est remplie, mais toi qui as encore droit d'espérer une longue carrière, ne la prodigue pas. J'ay cru m'apercevoir hier que tu étais triste ; pourquoi le serois-tu puisque je suis résigné tout et que je regarderai comme gagné tout ce que je ne perdrai pas. D'ailleurs nous ne sommes pas sans espérance de nous rejoindre, et en attendant, tes visites me font encore passer de doux instans."

Petitions from the commission of weights and measures and from the Committee on coinage for permission to Lavoisier to continue work in his laboratory were ineffectual, and soon afterwards the twenty-seven *fermiers généraux* were transferred to the hotel of the Ferme converted into a prison. Further efforts to save Lavoisier were vain; the terrorists having complete mastery of the Convention, accusations multiplied and defence became impotent. The unhappy prisoners were accused not only of plotting with the enemies of France, and of illegal exactions imposed on their fellow-citizens, but of mixing water with their tobacco and other ingredients hurtful to health.

Brought before Coffinhal, vice-president of the revolutionary tribunal, the prisoners were speedily and unanimously declared guilty by the jury. A defence put forward in the name of the *bureau de consultation* was put aside and it was probably then that Coffinhal pronounced the words, "La République n'a pas besoin de savans, il faut que la justice suive son cours."

The prisoners were taken back to the Conciergerie, and in a few hours the tumbrils were on their way to the Place de la Révolution. They were executed in the order in which their names stood on the act of indictment. Lavoisier, after seeing the end of his friend and father-in-law, Paulze, was the fourth to fall beneath the axe. Well might Lagrange say afterwards to Delambre: "A moment was all that was necessary in which to strike off this head, and probably a hundred years will not be sufficient to produce another like it."

The world has seen many revolutions since that day, and other revolutions will arise so long as injustice and oppression continue to be imposed by the few on the many. In these cases civilisation and all progress are delayed, and brute forces gain an ascendancy which leads to acts like those which disfigured the establishment of the first French Republic. England in permitting the persecution of Priestley by mob violence came very near a crime as great as the destruction of Lavoisier. In both countries these atrocious acts were followed by a revulsion of feeling, but too late to be of use to the victims concerned, for the one was dead and the other driven to end his days in a distant land.

One hundred and twenty-five years have elapsed since Lavoisier's death, and in the course of this long interval, by successive steps involving, beside the leadership of genius, the laborious experiments, calculation and reflection of thousands of men, chemistry has become one of the most complete and scientific branches of human knowledge. That position has been attained only by the consistent adoption of a principle enunciated by Lavoisier in his *Traité Elémentaire* (p. 52, *Œuvres*): "Il n'est jamais permis, en physique et en chimie, de *supposer* ce qu'on peut *determiner* par des expériences directes." It is this constant appeal to fact that renders the position of systematic chemistry unassailable. This does not, of course, preclude the use of hypothesis and what has been called the scientific use of the imagination, nor does it enable the most cautious thinkers to escape altogether the mistakes which arise from the temptation to enlarge even justifiable hypotheses. Lavoisier himself fell into one error of this kind when he was led to devise the name oxygen under the mistaken impression that this element enters into the composition of all acids without exception. Having adopted the definition of the word element, a definition originally given by Robert Boyle, as in chemistry meaning a simple body, the last product of analysis, Lavoisier showed that such bodies were capable of uniting in pairs to form compounds which in their turn were capable of combining together to form groups of a second order, and so a binary system was conceived by which for upwards of half a century the greater part of chemical phenomena were explained.

Group IV

ELECTRICITY IN THE SERVICE OF CHEMISTRY

Davy (1778-1829)

CHAPTER VII

DAVY

The end of the eighteenth century brought as great a change in the history of physical science as the French Revolution in relation to European affairs. The doctrine of phlogiston which had so long beclouded the views of chemists was completely dissipated by the new and clear teaching of Lavoisier, and the nomenclature of the new system with the names of the gases oxygen, hydrogen, azote, etc., was beginning to be familiar. Another great step in advance was nearly coincident with the beginning of the new century, for by that time the discoveries of Galvani and Volta were becoming familiar throughout the scientific world and the various forms of battery which succeeded the Voltaic Pile placed in the hands of experimenters a new and important instrument. With the writings of Black (1728-1799), of Cavendish (1731-1810), of Priestley (1733-1804) and of Scheele (1742-1786), the phraseology of the phlogistic period disappears. For although in the earlier papers of Lavoisier himself (1743-1794) the language of that system is necessarily used to some extent, the production of his *Traité de Chimie Elémentaire* in 1790 marks the final rejection of the old and adoption generally of the new order of things. With the new century began the career of one of the most attractive personalities and, among men of science, universally acknowledged geniuses, Humphry Davy. He has had many historians, for his immense popularity led to a kind of competition among biographers soon after his death. The two most famous lives of Davy are those by his brother, Dr. John Davy, and Dr. J. A. Paris. The former was written while on foreign service in Malta in 1832, but not published till three years later. The Life by Dr. Paris appeared in 1831 first in the form of detached sketches communicated to the weekly journal, the *Spectator*. The more complete biography

Sir Humphry Davy, P.R.S.
From the Portrait by Lonsdale.

[*Face p.* 78]

seems to have been undertaken with the consent of Lady Davy, though it could hardly have been with her approval, as it was already understood that the same task was contemplated by Dr. Davy. The consequence of this misunderstanding is that the world is supplied with the history of the same life from two points of view, very different in sentiment and in execution, and of which it may be said that the one was unnecessary and is in some respects inaccurate. To Dr. Davy's work one would naturally turn for information as being authentic, and though possibly prejudiced by brotherly affection, he gives a picture which has commended itself to successive generations by its manifest candour and directness. As to its completeness, it may be inferred that nothing of importance would be neglected or omitted which was known to the writer, who had all the materials in his hands and who wrote of his brother as "one who had acted the part of a father to me, whom I regarded as a brother, a teacher and most kind friend, and to whom I necessarily owed very much of what I most valued in life."

Humphry Davy was the son of Robert Davy and Grace Millet, his wife, and was born on December 17, 1778, in a house in Market Jew Street, Penzance. The family on both sides had been resident in the neighbourhood for some two hundred years and were small landowners. Robert Davy, the eldest son of Edmund Davy, a respectable builder, was brought up in the house of a great-uncle Robert, who died in 1774. Robert, having a great fondness for wood-carving, was allowed to go to London to obtain instruction, and became very proficient in the art. After his uncle's death, however, he did not permanently pursue the carving business, but after his marriage in 1776 he was occupied chiefly in farming a small property, called Varfell, in the parish of Ludgvan near Penzance, where he died in 1794. His widow was left in her thirty-fourth year with five children, of whom Humphry, the eldest, was then sixteen years of age. There were three daughters and the son John, who was born in 1790, and became his brother's biographer. In the early years of her widowhood Mrs. Davy's income was too small to enable her to educate and sustain her children, and she joined in a millinery business for three or four years till a small estate unexpectedly came into her possession and increased her income to about £300 a year.

The condition of society in Cornwall toward the end of the

eighteenth century was such as may be imagined when it is remembered that at that period Cornwall was still without great roads, and carriages were almost unknown. The usual mode of travelling was on horseback, and merchandise and news reached this far-away county by the same agency. Many of the Cornish people lived by smuggling, an occupation associated commonly with drunkenness and immorality. The demand and provision for education was narrowly limited, and the lower classes were extremely ignorant and superstitious. Davy's experiences at school were similar to those of other boys of his time, and he does not seem to have distinguished himself in the eyes of the schoolmaster. He was sent first to a Mr. Bushell, an old man who taught only reading and writing, but on the recommendation of this master he was removed at six years of age to the Grammar School.

The master of the Grammar School, the Rev. Mr. Coryton, was a man of irregular habits, and ill fitted for the office of teacher, generally indulgent as to the boys' work, but occasionally punishing slight offences with severity. He was addicted to pulling the boys' ears, and Davy, having suffered a good deal from this practice, appeared one day with a large plaster on each ear. When asked by the master what was the matter, he replied with a grave face that he had "put the plasters on to prevent a mortification."

Years afterwards, in a letter to his mother, he remarked: "After all, the way in which we are taught Latin and Greek does not much influence the important structure of our minds. I consider it fortunate that I was left much to myself when a child and put upon no particular plan of study, and that I enjoyed much idleness at Mr. Coryton's school. I perhaps owe to these circumstances the little talents that I have and their peculiar application. What I am I have made myself; I say this without vanity and in pure simplicity of heart."

At the age of fourteen he was removed to Dr. Cardew's school at Truro, where he remained during the year 1793. His school education was then considered complete, and on his return to Penzance he took up his abode with Mr. John Tonkin, a kind friend of the family, by whom he had been in a manner adopted and who had defrayed his expenses during the twelve months at Truro.

The greater part of the following year seems to have been

somewhat unsettled, and much time was given to fishing, shooting and aimless rambling. He began taking lessons in French with a refugee, M. Dugart, who was living in Penzance, but at the end of the year 1794 his father died and the event seems to have had the effect of settling his mind. In the beginning of 1795 he was apprenticed to Mr. Bingham Borlase, a surgeon and apothecary in Penzance, a man superior both in professional and general attainments to the majority of county practitioners. Davy received instruction probably as good as was possible in his day in the art and practice of his profession, but the note-books of the young student which have been preserved show that the plan of study he contemplated was conceived in the broadest spirit indicative of a mental activity far beyond the range which would be expected of a boy so young.

The earliest of the note-books, bearing the date 1795, contains the following scheme of study:

1. Theology.
    - Or Religion / Ethics or moral virtues — taught by Nature / by Revelation.
2. Geography.
3. My Profession.
    1. Botany.
    2. Pharmacy.
    3. Nosology.
    4. Anatomy.
    5. Surgery.
    6. Chemistry.
4. Logic.
5. Language.
    1. English.
    2. French.
    3. Latin.
    4. Greek.
    5. Italian.
    6. Spanish.
    7. Hebrew.
6. Physics.
    1. The doctrines and properties of natural bodies.
    2. Of the operations of nature.
    3. Of the doctrines of fluids.
    4. Of the properties of organised matter.
    5. Of the organisation of matter.
    6. Simple astronomy.
7. Mechanics.
8. Rhetoric and Oratory.
9. History and Chronology.
10. Mathematics.

Amid all this wide and discursive reading, which extended into history, poetry and the great dramatists, there was much writing of essays in which his own ideas are set forth and frequent outbursts in verse expressive of his own enjoyment and interest in the scenes of natural beauty found so abundantly in the country in which his boyhood was spent. From a child he had the reputation of being a rhymester, and when at school his best exercises were said to be translations from the classics into English verse, while his schoolfellows frequently resorted to him for help in the composition of their valentines and love-letters. In course of time some of his verses found publication in the *Annual Anthology*, and several poems at this early date justly entitle him to the epithet Poet and Philosopher, which one of his later biographers [1] has associated with his name. His first printed production bears the title *The Sons of Genius*, and is dated 1795. These and others of his poems breathe the joy which he always felt in scenes of natural beauty, and in his early years the view before him of Mounts Bay was a frequent source of inspiration. His travels in later life in other countries—France, Switzerland, Italy and the Tyrol—furnished themes of which he frequently made poetical use. Space, however, will not permit the introduction here of more than one or two examples of his verse in later years, and the reader interested in tracing the combination of poetry and philosophy in the mind of a great man of science will do well to read John Davy's Life of his brother, full as it is of charm on many sides; for it must be remembered that Davy was not only a chemist but a poet, not only a philosopher but a sportsman and a man of the world.

ASHBURNHAM PLACE,

JANUARY 22, 1823

Is this a time for minstrelsy,
When nature rests in deathlike sleep
And roots and buds and herbage lie
Embalmed in icy cerements deep ?

When scarce a stream is heard to flow
And scarce the distant woods appear,
So widely spreads the drifted snow,
The mantle of the new-born year ?

---

[1] *Humphry Davy, Poet and Philosopher*, by Sir Edward Thorpe, F.R.S. The "Century Science" Series (Cassells. 1896).

When the wild songsters of the grove
  Shivering around the mansion fly
Without a single note of love:
  Is this a time for minstrelsy?

It *is* a time for minstrelsy!
  For still the laurel blooms around
And bay, and Fancy's dreaming eye
  Can see through mists the fairy ground.

And hill and dale and woodlands green,
  And lakes which pastoral meads surround
The distant ocean and a scene
  At home where blossoms rise around.

And nature gains from art new powers,
  Charms that in happy union meet,
Where wild and cultivated flowers
  Together blend their odours sweet.

It *is* a time for minstrelsy,
  For round these walls what magic forms
Appear in grace and harmony!
  The pencil of the artist warms

The coldest scenes and powers sublime,
  Awakening moral forms of things,
And new creations steal from Time
  His scythe, and close his wings.

It *is* an hour for minstrelsy!
  For social converse wakes the mind
To pure and happy sympathy;
  And elegance and taste refined

Call to the hospitable board
  The force of reason and the flow
Of memory with wisdom stored
  Which might awake a grateful glow

In Fancy e'en though checked by age;
  Make sunshine in the darkest day,
And kindle in the coldest sage
  Some strain of vocal minstrelsy.

ON THE FALL OF THE TRAUN

July 25, 1827

From the high rock thy lovely waters burst
As if a new creation from the wand
Of Israel's mighty prophet, sprung to life
To save his people! But the dreamy thought
Of that most blessed, tho' but scanty rill
Gives but faint image of thy might and power
And awful force and fulness; as if a spirit
Imprisoned by magic art and now released
Thou thunderest on, determined to destroy;
And thy mild functions to produce and cheer
Are changed for attributes more terrible,
Saddening, destructive, wildly carrying on
Rocks, trees, before thee, e'en the mighty pine
Rending the mountain, through a new-torn vale
Opening thyself a passage to the plain.
But in thy wayward and most perilous leaps
Thou still art pure, and still might image well
The innate mind of poet or of sage.
In thy bright azure depths, and when thy foam
Sinks into quietness, I seem to view
That season of our life when pleasure fades
And sober reason with its heavenly light
Fills the cool deep of th' unimpassioned mind
Escaped from turbulent and fretful youth,
Its troubles, passions, bubbles, noise and foam
Which are well imaged in the falling stream.
E'en as I look upon thy mighty flood
Absorbed in thought, it seems that I become
A part of thee, and in thy thundering waves
My thoughts are lost, and pass to future time
Seeking the infinite, and rolling on
Towards the sea eternal and unbounded
Of the all-powerful, omnipresent mind!

Davy's note-books disclose the fact that he devoted practically an entire year to the study of elementary mathematics, beginning with "Fractions vulgar and decimal," by which one is reminded how little did this subject enter into the old-fashioned school course, arithmetic being taught as an extra by the writing-master in some of the great public schools till long after Davy's time. During the year 1797 his favourite exercise was metaphysics, and the notes show that he had made himself acquainted with the writings of Locke, Hartley, Bishop Berkeley, Hume,

Helvetius, Condorcet, Reid and his followers, beside Kant and the transcendentalists: a rather extensive field for a philosopher in his nineteenth year. This year also he began seriously the study of natural philosophy, including chemistry. His early chemical reading was confined to two works of very different character, namely, Lavoisier's *Elementary Chemistry* and Nicholson's *Dictionary of Chemistry*, the one distinguished for its admirable precision both as to fact and argument, the other an indifferent collection of facts and opinions hastily thrown together. He soon began experiments with the aid of such homely apparatus as he found at hand and with the common materials, such as mineral acids and alkalis furnished by the surgery. He made extraordinary progress, at least in the active conceptions of his own mind, for after the short space of four months he was corresponding with Dr. Beddoes of Bristol on the subject of what he called his Researches on Heat and Light. Unfortunately he was led to publish these essays a little later, but in the meantime they had the advantage of bringing him into communication with Dr. Beddoes, who conceived so high an opinion of the author as to offer him, before he had completed his medical studies, the charge of the patients in his Pneumatic Institution at Clifton.

During the last two years of his life at Penzance Davy had the advantage of the society of Gregory Watt, a son of the famous engineer, who came for the benefit of his health to the mild climate of Penzance and became a boarder in Mrs. Davy's house. He was a well-educated young man a little older than Davy, and interested in subjects of the same kind. Acquaintance was formed at about the same time with Davies Giddy, afterwards Gilbert (his successor in the Chair of the Royal Society), a man older than himself and possessing considerable scientific knowledge. Mr. Borlase seems to have conceived a high estimate of Davy's character, abilities and prospects, for he released his apprentice before the expiry of his indentures, on October 1, 1798. On the following day Davy left his home and, not yet twenty years of age, entered on a public career which in less than ten years made him one of the most famous men in Europe.

The Pneumatic Institution at Bristol, supported by subscription, was founded with the object of investigating the medicinal effects of the various gases which had been discovered or examined by chemists during the previous twenty years. Beddoes was described by Davy, in a letter to his mother, as "one of the most original men I ever saw—uncommonly short and fat, with little elegance of manners and nothing characteristic *externally* of genius or science; extremely silent, and, in a few words, a very bad companion."

The first undertaking by the young philosopher on his arrival at Clifton was the publication of a volume of essays, edited by Dr. Beddoes, which included an account of his own crude and inaccurate experiments on which he founded his theory of "phosoxygen," and the combinations of light. He was soon led to recognise the futility of theories erected on an insufficient foundation of fact, and probably the severe criticisms which he had, by undue haste, brought down on himself and his book had a salutary effect which extended through after years, and may be recognised in various expressions occurring in the papers in which the great discoveries of his more mature years are recorded. For example, in one of his papers on chlorine he remarks: "In the views that I have ventured to develope neither oxygen, chlorine nor fluorine are asserted to be elements; it is only asserted that, as yet, they have not been decomposed."

The first experimental enquiry undertaken at Clifton was the result of the observation that two pieces of cane rubbed together give out light, and that this phenomenon is connected with the presence of silica in the epidermis of grasses generally. Early in 1799, however, he was busy with the subject contemplated by the founders of the Institution, namely, the physiological effects of the various gases. One of the first tried was nitrous oxide, and so remarkable were the effects observed on its inhalation that the experiments recorded were sufficient to compose a volume published the following year. Nearly everyone is now familiar with the effects of inhaling nitrous oxide as an anæsthetic, and it can easily be imagined with what surprise and excitement the effects were first experienced. A few experiments were first made with the gas obtainable with the aid of zinc and dilute nitric acid, but after the discovery that nitrous oxide can be obtained in a state of purity by the application of heat to nitrate of ammonia, Davy determined to inhale the undiluted gas. He says: "I was aware of the danger of the experiment; it certainly never would have been made if the hypothesis of Dr. Mitchell had in the least influenced my mind." The hypothesis referred to assumed that the gas was "the principle of contagion when respired by animals in the minutest quantities." This Davy had proved by experiment to be groundless ; and the excitement and pleasurable sensations he experienced as a result of repeated experiments on himself led him to a characteristic effusion in verse :

> "Not in the ideal dreams of wild desire
> Have I beheld a rapture-wakening form :
> My bosom burns with no unhallowed fire,
> Yet is my cheek with rosy blushes warm ;
> Yet are my eyes with sparkling lustre fill'd ;
> Yet is my mouth replete with murmuring sound ;
> Yet are my limbs with inward transports fill'd
> And clad with new-born mightiness around."

He was led on by enthusiasm to more dangerous attempts to breathe marsh gas, carbonic acid gas, nitrogen, hydrogen and nitric oxide, from which he escaped with his life.

Throughout this time, as indicated by his note-books, he was actively occupied with his pen. Schemes and minutes of experiments were mixed up with lines of poetry, fragments of stories and romances, metaphysical fragments and sketches of philosophical essays involving notions which were recalled and recorded in some of the writings of his later years. Everywhere he displayed an ardent sympathy with nature and delight in the beauty of the world, associated with a conviction that "individuality can never cease to exist ; that ideal self which exists in dreams and reveries, that ideal self which never slumbers, is the child of immortality, and those deep intense feelings, which man sometimes perceives in the bosom of Nature and Deity, are presentiments of a more sublime and energetic state of existence." As he remarks elsewhere, "How different is the idea of life in a physiologist and a poet ! "

A letter to his mother, dated January 31, 1801, explains itself and shows the nature of the prospect which then lay before him, and which determined the character of his whole career. The following is the letter :

"My dear Mother,

"During the last three weeks I have been very much occupied by business of a serious nature. This has prevented me from writing to you, to my Aunt and to Kitty. I now catch a few moments only of leisure to inform you that I am exceedingly well, and that I have had proposals of a very flattering nature to induce me to leave the Pneumatic Institution for a permanent establishment in London. You have perhaps heard of the Royal Philosophical Institution, established by Count Rumford, and others of the aristocracy. It is a very splendid establishment, and wants only a combination of talents to make it eminently useful.

"Count Rumford has made proposals to me to settle myself there, with the present appointment of assistant-lecturer on chemistry and experimenter to the Institute; but this only to prepare the way for my being in a short time sole professor of chemistry, etc.; an appointment as honourable as any scientific appointment in the kingdom, with an income of at least 500*l.* a year.

"I write to-day to get the specific terms of the present appointment, when I shall determine whether I shall accept of it or not. Dr. Beddoes has honourably absolved me from all engagements at the Pneumatic Institution provided I choose to quit it. However, I have views here which I am loath to leave, unless for very great advantages.

"You will all, I dare say, be glad to see me getting amongst the *Royalists*, but I will accept of no appointment except upon the sacred terms of *independence*.

. . . . . .

"I am your most affectionate son,
"H. DAVY."

Little more than a month later he took up his abode in London, and in the spring of 1801, six weeks after his arrival, he gave his first lecture. He was then twenty-three years of age. His subject was the history and phenomena of "galvanism," then a very new and almost unexplored region. The Royal Institution was to be the scene of his later discoveries, and those of his immortal successor Michael Faraday.

Those who in the twentieth century have any experience of the evening lectures on Fridays at the Royal Institution cannot fail to have remarked the evidence of the scrupulous attention to detail in the preparations for the lecture, the punctuality of the lecturer and the rarity of any failure in experimental illustration; but they may not be aware that the importance and interest attaching to these Friday evening discourses is chiefly due to the example set by Davy and sustained by the long series of illustrious

LECTURE BY DR. GARNETT AT THE ROYAL INSTITUTION.
Davy acting as Assistant. Count Rumford standing by the door.
(*Gillray.*)

[*Face p.* 88]

professors, his successors at the Royal Institution. Nor can it be forgotten that the brilliant audience usually assembled has never been drawn together merely under the influence of a fashionable whim, for it is no less true now than it was in 1801 that the lecture theatre on these occasions is the habitual resort of many of the leaders in the scientific world.

It was with Davy an almost invariable rule to rehearse his lecture in the presence of his assistants the day before; this he did, not only with a view to the success of the experiments but also in regard to his own discourse. Dr. Davy records his recollection of the care with which his brother often repeated passages several times to study the effect of giving emphasis to particular words or expressions, and the result was an impressiveness which was the great charm of his oratory.

Davy's agreement with Count Rumford included the use of the laboratory and any apparatus required for new experiments, and on arriving in London he set to work immediately on some experiments already commenced in Bristol in connection with "galvanism," and published in *Nicholson's Journal* under the title "An account of some experiments made with the Galvanic Apparatus of Signor Volta."

The interior of a chemical laboratory in Davy's time was very different from what is to be seen in the present day. The laboratory at the Royal Institution was a large room well lighted from above and well supplied with water. The apparatus most conspicuous and most in use were a sand-bath for chemical purposes and for heating the room, a powerful blast furnace, a movable iron forge with double bellows, a blow-pipe apparatus attached to a table with bellows beneath, a large mercurial trough and two or three water-troughs and various galvanic troughs. It must be remembered that the last-named apparatus was designed to hold acid or saline solutions in which the copper and zinc plates were immersed when the current was wanted. Balances, air-pumps and other instruments liable to be damaged by acid fumes were kept in another room.

But Davy was not to be allowed to pursue his researches solely on lines laid down by himself. The Board of Managers were desirous of showing examples of the application of scientific principles to the arts, and accordingly the young professor was expected to give lectures on several subjects with which he could have no practical acquaintance. He seems, however, to have thrown himself quite readily into their plans, and in connection with the first subject proposed, namely, the art of tanning, he not only made experiments in the laboratory, but he visited tanyards and cultivated the acquaintance of practical tanners. The results of his enquiries were collected in several papers published in the *Journal of the Royal Institution* and the *Philosophical Transactions of the Royal Society* in 1802-3. But he had hardly entered on his investigations in the tanyard than he was called on by the Board of Agriculture to give a course of lectures on the connection of chemistry with vegetable physiology. Here he was not entirely a novice, for apart from the fact that he had learnt a little about practical farming from his father, he had given a good deal of attention to geology, and already in the Royal Institution itself he had given many lectures on this branch of science to the members. His native country had long before offered an attractive field of enquiry in this direction, with its variety of surface and the mines with which it is penetrated, from which he procured fascinating assortments of minerals. The time was also full of inducements to aid in the development of improved practices in agriculture, for a succession of bad harvests and the closure of continental ports by reason of the war had threatened famine. Davy's work brought him into contact with many important persons among land-owners and agriculturists, including the Duke of Bedford, who carried out experiments for him at Woburn, where trials have been systematically pursued down to the present day. Mr. Coke, of Holkham, Mr. Arthur Young and others were members of the Board of which Lord Carrington and Sir John Sinclair were successive Chairmen. The lectures were continued many years in succession, and in 1813 were published in the form of a quarto volume, for the copyright of which Davy received one thousand guineas with fifty guineas for each edition. It was certainly the most important treatise on agricultural science for nearly half a century till the subject was taken up again by Liebig.

Meantime Davy had not lost sight of the important subject of electro-chemistry, to which his attention was attracted immediately on the publication of the discoveries of Volta and the accidental observation in 1800 of the decomposition of water by means of the battery by Nicholson and Carlisle. A long series of experiments occupying several years was the result, and by 1806 he was able to advance the hypothesis "*that chemical and electrical attraction were produced by the same cause, acting in one case on particles, in the other on masses.*" But though this hypothesis was received with the most cordial approval by chemical authorities, Berzelius, the Swedish professor, among the number, Davy himself never attached great importance to hypotheses.

Pursuing his experiments, Davy was led to the discovery of the metallic elements potassium and sodium, and consequently the proof that the caustic alkalis, potash and soda, were no longer to be regarded as the simple or elementary substances they had hitherto been considered. Other methods of isolating these elements were afterwards devised by Gay-Lussac, but it is interesting to remember that the process of manufacture by which sodium is produced at the present day is essentially the same as that by which these elements were isolated in 1807-8, namely, by the electrolytic decomposition of caustic alkali in the fused state. This capital discovery at once placed Davy on the highest pinnacle of fame, and led to the further discovery that the alkaline earths, baryta, strontia and lime, like the alkalis, consist of oxides of metals possessing such remarkable characters that for a time it was even questioned whether they should be regarded as metals. The French chemists Gay-Lussac and Thénard regarded them as compounds of the alkalis with hydrogen, but this was never a view entertained by Davy. He was, however, not saved by his own researches from coming to an erroneous conclusion regarding ammonia, which he found to contain oxygen. In this case he was probably led astray in consequence of the general resemblance of ammonia in its combinations and reactions to the fixed alkalis, and his anxiety to show that oxygen was an essential constituent of such substances in opposition to the doctrine of Lavoisier, which regarded oxygen as an acidifying principle.

Towards the end of 1807 Davy was struck down by severe illness, which there can be little doubt was caused by the over-fatigue and excitement connected with his experimental labours and the resulting discoveries, acting on an ardent and enthusiastic nature.

In illustration of this it was reported by Mr. Edmund Davy, his cousin and assistant, that when he saw the minute globules of potassium burst through the crust of potash and take fire he could not contain his joy; he actually danced about the room in ecstatic delight, and it was some time before he was calm enough to continue the experiment. His illness lasted nine or ten weeks, and, needless to say, it excited the most anxious sympathy among his friends and among the members of the institution and the public. During his convalescence he turned characteristically to the expression of his thoughts in verse, and completed a poem already begun, to which he now gave the heading: "Written after recovery from a dangerous illness." It is too long to quote entire, and a single verse of the eighteen of which it is composed would give an imperfect and erroneous impression of the beauty of the lines, which trace the course of human existence illustrated by such a life as his own, and are expressive of his own visions of immortality.

His researches in the constitution of the alkalis and earths were communicated to the Royal Society in four successive Bakerian Lectures.

He was next engaged in an enquiry almost equal in importance, namely, the true nature of the gas originally discovered by Scheele and called by him *dephlogisticated marine acid air*, but later, in accordance with Lavoisier's system, *oxymuriatic acid*. The question was whether this gas contained oxygen, as implied by the name. To determine this point Davy made the experiment of heating in the gas a variety of substances distinguished for the readiness with which they combine with oxygen, such as potassium, and showing that the resulting compounds are not oxides. Carbon strongly ignited in the gas undergoes no change. At the conclusion of all his work he established the view ever since held concerning the nature of the gas as an elementary, or at any rate undecomposable substance, and he gave to it the name *chloric gas* or *chlorine*, in reference to its colour, a name which implies no theory and would not require to be changed even if it were discovered hereafter to be a compound.

Other researches relating to fluoric compounds did not enable Davy to isolate the peculiar element contained in them, but he succeeded in satisfying his mind as to its existence and its analogy with chlorine sufficiently to justify him in assigning to it the name *fluorine*.

The discovery of *euchlorine*, of hydrogen telluride and hydrogen phosphide, and several other compounds of greater or less interest and importance, followed about this time. The interest roused by his great discoveries of the alkali metals led to the formation of a fund by which a great voltaic battery of 2,000 plates was constructed for his use in 1809. With this battery he obtained some remarkable effects, among which was the production of a vivid light when charcoal points attached to the terminal wires of the battery were brought into contact and then slightly withdrawn from each other. It is especially interesting to us, accustomed to the use of electric light in various forms, to recollect that the arc was first beheld by the human eye in Davy's lecture room at the Royal Institution. The effects were described by him in the following words: "When pieces of charcoal about an inch long and one-sixth of an inch in diameter were brought near each other (within the thirtieth or fortieth part of an inch) a bright spark was produced, and more than half the volume of the charcoal became ignited to whiteness, and by withdrawing the points from each other a constant discharge took place through the heated air in a space equal at least to four inches, producing a most brilliant arch of light, broad and conical in form in the middle. . . . When the communication between the points positively and negatively electrified was made in air rarefied in the receiver of the air-pump, the distance at which the discharge took place increased as the exhaustion was made; and when the atmosphere in the vessel supported only one-fourth of an inch of mercury in the barometrical gauge, the sparks passed through a space of nearly half an inch, and by withdrawing the points from each other the discharge was made through six or seven inches, producing a most beautiful coruscation of purple light," etc.

Toward the end of 1811 Davy made the acquaintance of the lady who soon afterwards became his wife. She was the widow of Shuckburgh Ashby Apreece, the eldest son of Sir Thomas Apreece, and, daughter and heiress of Charles Kerr of Kelso. She had been a prominent figure in society both in Rome and in London, and, according to Sir Walter Scott, of whom she was a distant relation, she had been "a lioness of the first magnitude in Edinburgh." The marriage took place on April 11, 1812, Davy having been knighted by the Prince Regent only three days before. "This distinction," he remarked in a letter to his brother, "has not often been bestowed on scientific men, but I am proud of it, as the greatest of human geniuses bore it, and it is at least a proof that the Court has not overlooked my humble efforts in the cause of science."

It is a curious indication of the position of science in the country at this time that the only honour conferred on Davy by a University was the LL.D. given to him at Trinity College, Dublin, on the occasion of a course of lectures he delivered there in 1811. This was, of course, a consequence of the state into which the Universities had fallen, or rather from which they had never emerged, and which was the subject of so much pungent and deserved satire from the pen of the Rev. Sydney Smith.

The honeymoon appears to have been spent at Beechwood Park, near St. Albans, and in a round of visits in the north of England and in Scotland.

In the meantime Sir Humphry had notified the managers of the Royal Institution that he could no longer continue to lecture, and on June 1st he was appointed Honorary Professor of Chemistry and Director of the Laboratory. He was succeeded as Professor of Chemistry by William Thomas Brande, then a very young man, who retained this office for many years. A more important event which occurred before Davy's lectures came to an end was the introduction of Michael Faraday. The boy had obtained access to the lectures, and having taken careful notes he sent a copy with illustrative drawings of apparatus to Sir Humphry, who acknowledged their merits, and in March, 1813, secured for the applicant the post of assistant in the laboratory. In October of the same year he accompanied Sir Humphry and Lady Davy as secretary and assistant in experiment in their travels on the continent of Europe.

It must not be forgotten that England was then at war with France, but his scientific fame secured for Davy not only permission from the French Government to visit the country, but a warm and friendly welcome from the scientific men in Paris. He was even elected a Corresponding Member of the Institute of France and was awarded the Napoleon Prize of the Institute. The travellers crossed from Plymouth to Morlaix in a "cartel" and were a day and two nights at sea. They spent about two months in Paris, where, having "commodious portable apparatus with him," Davy occupied himself partly in the society of the numerous scientific men of that city and partly in research. Of the former he left notices of a few of the most distinguished, among them Guyton de Morveau and Vauquelin, both associates of Lavoisier and now in the decline of life. Guyton de Morveau he describes as an old man of mild and conciliatory manners, though in the time of the Revolution he had been a violent republican. Vauquelin gave the idea of a man of a past age belonging rather to the pharmaceutical than to the philosophical world. "Nothing could be more singular than his manners, his life, and his ménage. Two old maiden ladies, the Mademoiselles de Fourcroy, sisters of the Professor of that name, kept his house. "I remember," he says, "the first time that I entered it I was ushered into a sort of bedchamber, which likewise served as a drawing-room. One of these ladies was in bed, but employed in preparations for the kitchen, and was actually paring truffles. Vauquelin wished some immediately to be dressed for my breakfast, and I had some difficulty to prevent it. Nothing could be more extraordinary than the simplicity of his conversation; he had not the slightest tact and, even in the presence of young ladies, talked of subjects which, since the paradisaical times, never have been the objects of common conversation."

"Gay-Lussac was quick, lively, ingenious and profound, with great activity of mind, and great facility of manipulation. I should place him at the head of the living chemists of France.

"Berthollet was a most amiable man; when the friend of Napoleon even, always good, conciliatory and modest, frank and candid. He had no airs and many graces. In every way below la Place in intellectual powers, he appeared superior to him in moral qualities. Berthollet had no appearance of a man of genius; but one could not look on la Place's physiognomy without being convinced that he was a very extraordinary man."

Beside the chemists he met many eminent men representative of other branches of science. Of Cuvier he wrote: "I should say of him that he is the most distinguished man of *talents* I have known, but I doubt if he is entitled to the appellation of a man of genius." And of Humbolt that he was "one of the most agreeable men I have ever known. . . . His works are monuments of the variety of his knowledge and resources."

Davy engaged immediately in the experimental examination of the remarkable substance discovered about two years earlier by Courtois, a manufacturer of saltpetre, and to which the name *iodine* had been given in reference to the colour of its vapour. Davy received a small quantity from his friend Ampère and soon proved that the acid formed from it was distinct from muriatic acid, and that iodine itself was a simple substance analogous to chlorine.

On December 23rd he left Paris for the south of France. Of course he composed much verse on his way: at Fontainebleau on December 29th; at Lyons, on a distant view of Mont Blanc, January 5th; on the banks of the Rhone, January 6th; on the Mediterranean Pine at Montpelier, January 14th; on the Canigou, one of the highest points of the Eastern Pyrenees, January 26th; at Vaucluse in the beginning of February, 1814, and at other points. He crossed into Italy by way of Nice and the Col di Tenda, and passing through Turin he proceeded to Genoa, where he took the opportunity of making some experiments on the electricity of the torpedo. In the middle of March the travellers reached Florence, where Davy made some experiments on the combustion of the diamond in oxygen, using the great lens in the cabinet of the Academy. He found that the diamond, once ignited, continued to glow in oxygen gas, but without producing any alteration in the volume of the gas, which consisted of pure carbonic acid, no water being formed.

In the beginning of April the Davys left Florence for Rome, where they remained nearly a month. They then went on to Naples and returned to Rome in the last week of May, passing on almost immediately with the intention of spending the summer in Switzerland.

Between Rome and Naples they met with a party of brigands in the passes of the Apennines, but after an amusing conversation with the captain of the party in walking together up a steep ascent, they were allowed to pass unmolested. In May he was present at the return from banishment of the Pope Pius VII. and his entry into Rome borne on the shoulders of the most distinguished artists, headed by Canova, whose acquaintance Davy made and to whom he addressed a short poem.

In passing through Milan Davy met Volta, then nearly seventy years of age, and in bad health. In his notes of distinguished men, Davy says of Volta that "his conversation was not brilliant; his views rather limited, but marking great ingenuity. His manners were perfectly simple. He had not the manner of a courtier or even of a man who had seen the world."

From Milan he crossed the Alps by the Simplon and arrived in Geneva at the end of June. Here the pair remained about three months enjoying the best society, chiefly English, and the pleasure of fishing, always Davy's favourite sport. Returning to winter in Italy they visited parts of the Tyrol, concerning the beauty of the scenery and the interesting geology of which he left some notes. The whole of the winter was spent in Rome. Here he made some analyses of the pigments employed in painting by the ancients, and sent home to the Royal Society some experiments on iodic acid and the perchlorates. From Rome he visited Naples, ascended Vesuvius and went round the base of Somma. By the time spring was well advanced Davy and his wife made their way back to England, avoiding France by making a detour through part of Germany and Flanders. On May 5, 1815, he wrote to his mother from a London hotel: "We have had a very agreeable and instructive journey, and Lady Davy agrees with me in thinking that England is the only country to *live* in, however interesting it may be to *see* other countries. I yesterday bought a good house in Grosvenor Street, and we shall sit down in this happy land."

But there was another person who was doubtless glad to be back in his own country again, and that was the young assistant Faraday, who had accompanied Sir Humphry and Lady Davy in all their journeyings. For we have the testimony of Faraday himself that he often found his position as travelling secretary and assistant unpleasant owing to the whims and occasional exhibitions of irascible temper on the part of the lady. It was unfortunate that Davy's valet, who had arranged to travel with him, was afraid of trusting himself in the enemy's country and ultimately declined to go. A servant could not be found in all the continental towns they visited, and the consequence was that Faraday was sometimes placed in a somewhat humiliating position. Lady Davy has been described as "small, with black eyes and hair and a very pleasant face, an uncommonly sweet smile, and when she speaks has much spirit and expression in her countenance." She had been accustomed to play a leading part in society, and perhaps the attention paid to her illustrious husband may have tended to diminish the brilliancy of her own position. It is, at any rate, a question whether she was in all respects fitted to be the wife of such a man, and it is significant that in his later years and when in bad health he wandered about alone on the Continent, and, as his brother says, "dependent on his own resources; no friend to converse with; no one with him to rely on for aid, and in a foreign country without even a medical adviser; destitute of all the amusements of society; without any of the comforts of home—month after month he kept on his course wandering from river to river, from one mountain lake and valley to another in search of favourable climate; amusing himself with his gun and his rod when sufficiently strong to use them, with 'speranza' for his rallying word." Lady Davy was with him only during his last days. She lived a widow till 1855.

Soon after his return to London in 1815 Davy entered on the train of researches on the structure and conditions of flame which resulted in the invention of the "Safety Lamp" for miners.

In evidence of the importance and value of the discovery of the principle and its application to the purposes of the miner, a service of plate valued at about £1,200 subscribed for by the chief colliery owners of the north of England was presented to Sir Humphry on October 11, 1817. Beside the present from the coal-owners he also received a silver-gilt vase from the Emperor Alexander of Russia, and a baronetcy was conferred on him in 1818.

On May 26th Sir Humphry and Lady Davy started for Vienna, in which city they arrived about June 13th. After passing through part of Hungary they made various excursions in Styria, Idria and Carniola. He visited the salt mine at Hallstadt and in a letter to Professor Brande he mentioned that they were still ignorant of the use of the Safety Lamp, in which he gave them instructions. "The inflammable air," he says, "is found in largest quantities where the *blue* salt is. I have been again searching in vain for the cause of this extraordinary colour." To this day this phenomenon has not been fully explained, though it has been attributed to the action of radio-active matter, and the liberation of the alkali metal.

By the end of October Davy was in Naples making experiments on the rolls of papyri from Herculaneum. In this he was only partially successful, and the fragments disclosed possessed little literary interest. Another subject in which he occupied himself was in observation of the phenomena of volcanoes, concerning which he had theories of his own.

The Davys left Naples early in the spring of 1820, travelling slowly, as Lady Davy was reported not very well, and reached London on June 20th.

On June 19th Sir Joseph Banks died, having been President of the Royal Society during forty-two years. Davy immediately offered himself, and on November 30th, St. Andrew's Day, 1820, when the officers are annually elected, the choice of the Fellows was almost unanimously in his favour.

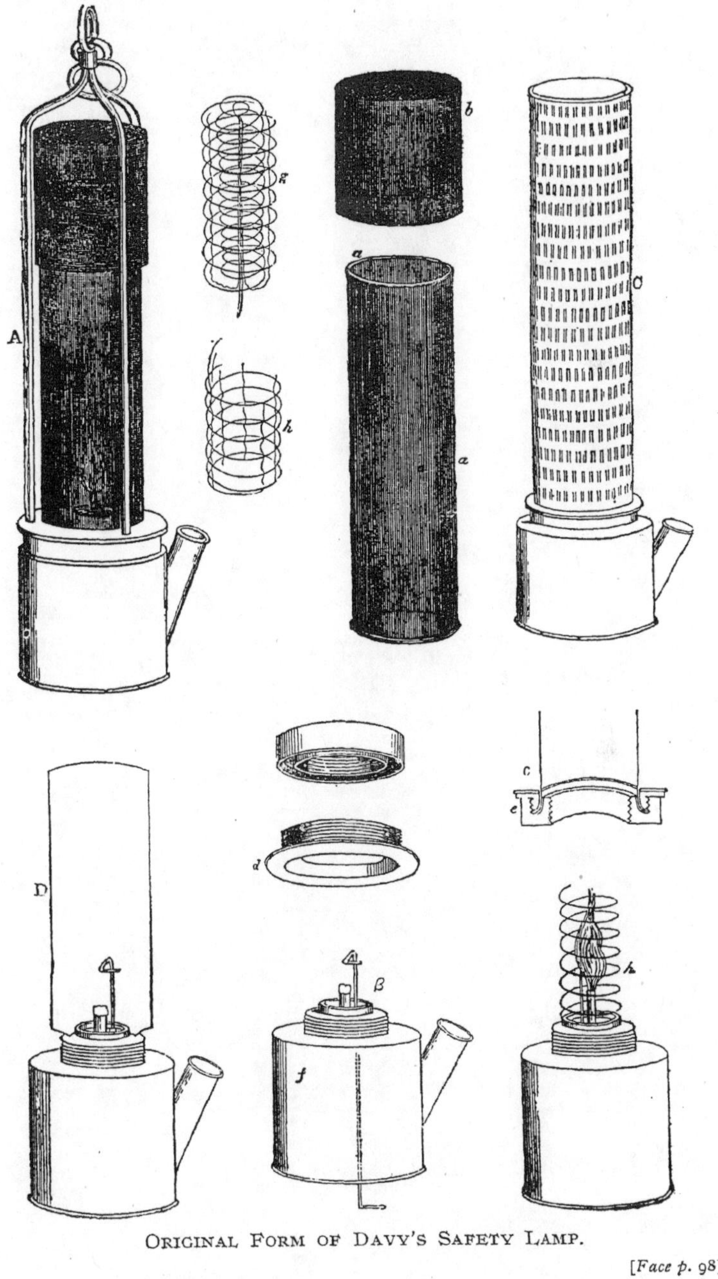

ORIGINAL FORM OF DAVY'S SAFETY LAMP.

[*Face p.* 98]

Of Sir Joseph Banks, his predecessor, it was no doubt true that, according to Davy's estimate, "he was a good-humoured and liberal man, free and various in conversational power, a tolerable botanist and generally acquainted with natural history. He had not much reading, and no profound information. He was always ready to promote the objects of men of science, but he required to be regarded as a patron, and readily swallowed gross flattery." Davy was a man of very different character, habits and ideas. As it had been the custom of former Presidents to observe a certain state in all that related to their office, Davy followed their example and continued to take the chair in Court dress, sitting covered, with the mace presented by the Royal Founder before him. So long as he lived at 28, Lower Grosvenor Street, he continued to give evening parties on Saturdays, which were attended by many men of science beside artists, literary men and others. Afterwards, on removing to 26, Park Street, Grosvenor Square, these gatherings were discontinued, but the library of the Society then at Somerset House was opened on Thursday evenings after the regular meetings were concluded.

Davy commenced his presidential duties with a high sense of their importance and with many schemes for improving the reputation of the Society and increasing its usefulness. But he soon found that the worries inseparable from this kind of office, involving not only attention to the details of the regular business of the Society, but the necessity for sometimes rejecting papers and otherwise interfering with the projects of people desirous of making use of the Society for their own purposes, caused him much anxiety and annoyance. Davy was re-elected to the presidential chair seven years in succession, but though he continued to make researches on a variety of subjects, partly in the laboratory of the Royal Institution, and whenever opportunity offered during his travels on the Continent, the day of great discovery was for him past.

Researches on electro-magnetism, discovered by the Danish philosopher Œrsted, and developed in later years by Faraday, was one of the most important subjects he attacked.

A subject of great practical importance was brought to his notice by the Commissioners of the Navy on account of the serious loss incurred in consequence of the corrosion of the copper sheathing with which the ships of that day were usually covered. Davy's discovery of a means of protection by attaching to the copper a relatively small area of a more electro-positive metal, such as zinc or iron, promised at first to be a complete cure for the trouble. For a variety of reasons, however, especially the tendency for seaweeds and shell-fish to accumulate on the protected surface, the method was ultimately abandoned. In the summer of 1824 he made a voyage to the North Sea with the object of trying the influence of motion on the protectors. He was absent some weeks travelling round the coast of Norway, through Sweden, Denmark, Holstein and Hanover, sometimes in the Admiralty steamboat and sometimes by land.

Of course he carried everywhere his gun and rod, and according to his diaries of the journey he had good sport, especially on the lakes and rivers. He dined with the Crown Prince of Sweden and the Princess, a granddaughter of the Empress Josephine; he also visited Œrsted, Berzelius, Olbers, the astronomer, and Gauss, the mathematician and physicist, and others. Of some of these distinguished men he left sketches of character. "*Berzelius*," he says, "was the worthy countryman of Scheele, and certainly one of the great ornaments of the age. Indefatigable in labour, accurate in manipulation, no one has worked with more profit. His manner was not distinguished, his appearance rather coarse, and his conversation limited much to his own subjects.

"*Œrsted* is chiefly distinguished by his discovery of electro-magnetism. He was a man of simple manners, of no pretensions, and not of extensive resources, but ingenious, and a little of a German metaphysician."

In September, 1826, Davy's mother died, and soon after his own health began to decline. In December he had a paralytic attack from which, however, he recovered sufficiently to be able to undertake a journey to the Continent in charge of his brother, who was an army surgeon. The winter was very severe and the whole of France covered with snow. After a difficult crossing of Mont Cenis they found much snow throughout Lombardy, and it was only on arriving at Ravenna in March that it began to disappear. By the kindness of the Vice-Legate of Ravenna he was lodged in the Apostolical Palace. Here he was alone, but was well enough to ride in the forest. But the melancholy attitude of his mind was indicated by the verse which from time to time fell from his pen, of which an example is shown in the lines dated Ravenna, April, 1827:

"Oh, couldst thou be with me, daughter of heaven,
Urania! I have now no other love;
For time has withered all the beauteous flowers
That once adorned my youthful coronet."

The reading he then preferred was Byron's poems, and he found additional interest in Ravenna, where he met the Countess Guiccioli and her family. Occupying himself, as much as was possible to an invalid, in natural history observations and in experiment, he passed northward on the approach of summer and remained in Switzerland and the Austrian Tyrol till, at the end of June, being at Salzburg, and in grave doubt of permanent recovery, he wrote to Mr. Davies Gilbert resigning office as President of the Royal Society. With all the care possible his health did not improve, and he determined to return to England, where he landed on October 6th.

He remained in England till the following March, and being unable to follow actively his ordinary pursuits either in science or sport, he sought refuge from ennui in composing his work on fly-fishing, which he called *Salmonia*. All his life he had been devoted to angling, while after his marriage he seems to have preferred the gun, possibly owing to more frequent opportunities. *Salmonia*, like the *Complete Angler* of Isaac Walton, is in the form of a dialogue, but it is not confined to the subject indicated by the title. His brother and biographer remarks that "never was a work more characteristic of the mind and pursuits of its author," including as it does an account of his experience in angling and his researches in natural history and science, diversified and enriched by thoughts and reflections rising from appreciation of the beauties of nature. It is the work of the sportsman-poet or poet-sportsman, and is even at the present day familiar to kindred spirits.

As his health did not improve, Davy set out on March 29, 1828, with Dr. Tobin, the eldest son of his early friend, Mr. James Tobin, of Bristol, with the intention of passing the summer in his favourite districts of Southern Austria and descending into Italy on the approach of winter. While on his journey he planned and in part composed his last work, *Consolations in Travel, or The Last Days of a Philosopher*. This, like *Salmonia*, consists of a series of dialogues in which one of the characters may be considered to represent himself in ordinary life, while another may be supposed to represent what he wished to be as a chemist, but more especially as a philosopher.

Arriving at Laybach on May 4th, he passed on to Ischl and Salzburg, and in October he went to Trieste expressly for the purpose of trying long meditated experiments on the electricity of the torpedo. The results were sent to the Royal Society in a paper which was printed in the *Philosophical Transactions* for 1829. It was his last contribution, but experiments were continued after his death by his brother John. In November he was again in Rome, but towards the end of February another paralytic seizure occurred, and Dr. Davy, who was in Malta, hurried to him. Lady Davy also arrived from England, bringing a copy of the second edition of *Salmonia*, which gave him great pleasure. His condition improved sufficiently to allow him at the end of April to undertake the journey to a cooler climate, and by easy stages the party reached Geneva on May 28th. He bore the journey well, but on hearing from Lady Davy of the death of his old friend, Dr. Thomas Young, he was deeply affected; and he was further upset by a slight accident, striking his elbow against the arm of the sofa on which he sat. He was got to bed as quickly as possible, but the end came about 3 o'clock in the morning of May 29th.

His remains were deposited in the cemetery of the city and close to the grave of Professor Pictet, the funeral being attended by the Council of State and representatives of the Academy. An obelisk with a Latin inscription marks the spot.

Thus passed away, at the early age of fifty-one, one of the most remarkable men of this or any other time. The precocious genius which could lift an almost uneducated boy from the condition of a surgeon's unqualified apprentice to a position among the greatest scientific discoverers in the short space of ten years, was combined with a poetic fervour which alone would have ranked him among the remembered poets of his day. The friend of Coleridge, of Southey, of Wordsworth and Sir Walter Scott, every one of whom left their testimony to his powers, he was at the same time endowed with the most brilliant conversational powers, which are described, for example, in connection with his visit to Abbotsford in Lockhart's Life of Scott. The estimate formed of his scientific position is well represented in an article in *Silliman's American Journal* immediately after his death. The writer says: "Some might even entertain

DAVY'S TOMB IN THE CEMETERY AT GENEVA.

[*Face p.* 102]

the apprehension that so extensive a popularity among his contemporaries is the presage of a short-lived fame; but his reputation is too intimately associated with the eternal laws of nature to suffer decay; and the name of Davy, like those of Archimedes, Galileo and Newton, which grow greener by time, will descend to the latest posterity."

Group V

LAWS OF COMBINATION AND THE ATOMIC THEORY

DALTON (1766-1844) PROUST (1755-1826)
GAY-LUSSAC (1778-1850) BERZELIUS (1779-1848)

CHAPTER VIII

DALTON

CONCERNING Davy's contemporaries, Priestley, who died in 1804 in a distant land, and Cavendish, who died in 1810, something has already been said, but there is another, his senior by some twelve years, who exercised on the progress of science an influence as great though in quite a different direction. This was JOHN DALTON, whose name has been for more than a century associated with the application of the Atomic Theory to chemistry. The Atomic Theory assumes that the established laws of combination between the elements are accounted for by the hypothesis that matter consists of separate particles limited in size and weight, and that a mass of any given element, such as oxygen or iron, consists of particles all alike in every respect. When a compound is formed the particles of the compound are also all alike, and, using Dalton's own words, "Every particle of water is like every other particle of water; every particle of hydrogen is like every other particle of hydrogen, etc. When any body exists in the elastic state, its ultimate particles are separated from each other to a much greater distance than in any other state; each particle occupies the centre of a comparatively large sphere, and supports its dignity by keeping all the rest, which by their gravity, or otherwise, are disposed to encroach upon it, at a respectful distance. . . . Chemical analysis and synthesis go no farther than to the separation of particles one from another, and to their reunion. No new creation or destruction of matter is within the reach of chemical agency. We might as well attempt to introduce a new planet into the solar system, or to annihilate one already in existence, as to create or destroy a particle of hydrogen. All the changes we can produce

John Dalton.

From Engraving by C. Turner, A.R.A., of Picture by J. Lonsdale.

[*Face p.* 104]

boy of twelve, at the Eaglesfield school. Sometimes they challenged him to fight in the adjoining graveyard, and they are reported to have broken the school windows when he locked them in to complete their tasks at dinner time. But sharing afterwards in the labour on his father's little farm, he grew up strong and healthy in body and mind. When he joined his brother Jonathan at Kendal the school there was not generally popular, and according to the recollections of a lady who was at school under the brothers in 1785, this was partly due to the uncouth manners of the young masters.

During this period John contributed frequently to two periodicals, *The Gentlemen's Diary* and *The Ladies' Diary*, and on more than one occasion received a prize for solution of problems in mathematics and philosophy. This period of his life was also memorable from his close acquaintance with Mr. John Gough, the son of a tradesman in Kendal who had been rendered quite blind by smallpox when about two years old and was then about thirty. Dalton, in the preface to the first edition of his *Meteorological Observations and Essays*, 1793, expressed his sense of obligation in very warm terms. And indeed Gough must have been, in Dalton's words, "one of the most astonishing instances that ever appeared of what genius united with perseverance and every other subsidiary aid can accomplish, when deprived of what we usually reckon the most valuable sense." Gough was not only a good classical scholar and mathematician, but had an extensive knowledge of the science of the day. Dalton states that "it was he who first set the example of keeping a meteorological journal at Kendal." Dalton himself commenced one entitled *Observations on the Weather, etc., etc., March* 24, 1787. This is a matter of some historical importance, because Dalton's earliest scientific observations, continued through many years, led him, it is supposed, to those considerations regarding the constitution of gases which culminated in the Atomic Theory a few years later. Dalton's instruments used in this work were constructed by himself. They were doubtless inaccurate enough according to modern standards, but they served his own purpose and were considered satisfactory by other people, for he supplied to Mr. Peter Crosthwaite, the founder of the Keswick Museum, a barometer and a thermometer for which he charged him 18s. and 5s., and two thermometers were sent to his early friend, Elihu Robinson, to whom also he communicated the details of their construction. The inconveniences and difficulties surrounding such operations in those days may be imagined from the fact that in order to heat the mercury a candle flame was applied to the bulb. It will be remembered that the candle was a common source of heat used by many of the older experimenters—Priestley, for example.

In 1787 Dalton tried his hand at public lecturing, and in October of that year he gave a course of twelve lectures on "Natural Philosophy," "Subscribers to the whole, half a guinea; or one shilling for single nights." These lectures were repeated in after years at Kendal and elsewhere, with, however, scanty success. He was never a good lecturer, and his experiments were often unsuccessful even in later years, when he had acquired greater experience. While in Kendal he made a fairly complete collection of the plants of the district, and of the butterflies and moths, which was deposited in the Keswick Museum. He began to think of undertaking the study of medicine, with a view to practice, but was dissuaded from the attempt by his want of means and by the representations of his friends. In 1793 he left Keswick and went to Manchester, to fill the position of tutor in mathematics and natural philosophy at New College. This college had been established in 1786 by the Presbyterians as a continuation of the Warrington Academy with which Priestley and others had been honourably connected. This was the forerunner of the Manchester College ultimately erected in Oxford as a protest against the exclusiveness of Oxford and Cambridge prior to 1871, when religious tests were abolished. Dalton seems to have been happy in his new circumstances. In a letter to Elihu Robinson in 1794, after describing the college building and his own room on the top floor, which "has two windows and a fireplace, is handsomely papered, light, airy and retired," he proceeds: "whether it is that philosophers like to approach as near to the stars as they can, or that they choose to soar above the vulgar into a purer region of the atmosphere I know not; but my apartment is full ten yards above the surface of the earth. . . . There is in this town a large library furnished with the best books in every art, science and language, which is open to all gratis; when thou art apprised of this and suchlike circumstances, thou considerest me in my private apartments, undisturbed, having a good fire and a philosophical apparatus around me, thou wilt be able to form an opinion whether I spend my time in slothful inactivity of body and mind. The watchword for my retiring to rest is 'Past twelve o'clock—cloudy morning.'

"There is a considerable body of Friends (Quakers) here; near 200 attend our first day (Sunday) meetings. I have received particular civility from most of them, and am often at a loss where to drink tea on a first-day afternoon, being pressed on so many hands."

In December, 1797, he wrote to his brother that his time was so much taken up with tuition at home and in the town together that he can scarcely turn to any mathematical or philosophical pursuit. And so as time went on he began to consider the advantages to his freedom which would result from giving up the tutorship at college and gaining his living as a private tutor. Accordingly in 1799 he retired from the post and henceforward devoted his life to scientific work, but earning his bread as a private teacher, of mathematics and natural philosophy chiefly, together with English, in connection with which subject he published a Grammar in 1801. His *Observations* in meteorology had been published, as already mentioned, in 1793, and the first entry made in 1787 described a remarkable display of aurora, a phenomenon which interested him very much. He cannot be said to have discovered the connection between the display and terrestrial magnetism, for this had already been indicated by the astronomer Halley nearly a century earlier. But Dalton, with characteristic independence, paid but little attention to the work of others or to their opinions when they did not accord with his own. During his meteorological observations, extending over several years, he was meditating in his own fashion the properties of gases and endeavouring to account for them; and that it was in connection with these observations that the Atomic Theory assumed a definite shape in his mind appears now to be certain, from the discovery of manuscript notes of his in the collection of his papers preserved by the Manchester Literary and Philosophical Society. In the winter 1803-4 he gave a course of lectures at the Royal Institution in London, and in a second course at the Royal Institution, given in 1809-10 he explained exactly and in considerable detail how he had been led to the doctrine of particles. A few extracts from these notes may be given here, but the reader must be referred for the complete story to the *Essay* of Sir Henry Roscoe and Dr. Arthur Harden published in 1896. In the notes prepared for the seventeenth lecture Dalton gives an account of his recollections as follows: "Having been long accustomed to make meteorological observations and to speculate upon the nature and constitution of the atmosphere, it often struck me with wonder how a *compound* atmosphere or a mixture of two or more elastic fluids should constitute apparently a homogeneous mass, or one in all mechanical relations agreeing with a simple atmosphere.

"Newton had demonstrated clearly in the 23rd Prop. of Book II of the Principia that an elastic fluid is constituted of small particles or atoms of matter which repel each other by a force increasing in proportion as their distance diminishes. But modern discoveries having ascertained that the atmosphere contains three or more elastic fluids of different specific gravities, it did not appear to me how this proposition of Newton's would apply to a case of which he, of course, could have no idea. The same difficulty occurred to Dr. Priestley, who discovered this compound nature of the atmosphere. He could not conceive why the oxygen gas, being specifically heaviest, should not form a distinct *stratum* of air at the bottom of the atmosphere, and the azotic gas at the top of the atmosphere. Some chemists upon the Continent—I believe the French—found a solution of this difficulty (as they apprehended). It was *chemical affinity*. One species of gas was held in solution by the other, and this compound in its turn dissolved water—hence evaporation, rain, etc. . . . It was objected that there were no decisive marks of chemical union when one kind of air was mixed with another. The answer was that the affinity was of a very slight kind, not of that energetic cast which is observable in most other cases. I may add, by the bye, that this is now, or has been till lately, I believe, the prevailng doctrine in most of the chemical schools in Europe."

He then explains how he endeavoured to reconcile these ideas by assuming each atom surrounded by an atmosphere of heat and making them respectively centres of repulsion, whether in a mixed or simple state. There were, however, still difficulties, and in considering the subject further "it occurred to me that I had never contemplated the effect of *difference of size* in the particles of elastic fluids. . . . This idea occurred to me in 1803." "The different *sizes* of the particles of elastic fluids under like circumstances of temperature and pressure being once established, it became an object to determine the relative *sizes* and *weights*, together with the relative *numbers* of atoms in a given volume. This led the way to the combinations of gases and to the number of atoms entering into such combinations." In the next lecture Dalton explained that he had "chosen the word *atom* to signify these ultimate particles in preference to *particle, molecule* or any other diminutive term, because I conceive it much more expressive; it includes in itself the notion of indivisible which the other terms do not. It may perhaps be said that I extend the application of it too far when I speak of *compound atoms;* for instance, I call an ultimate particle of *carbonic acid* a *compound atom.* Now though this atom may be divided, yet it ceases to become carbonic acid, being resolved by such division into charcoal and oxygen. Hence I conceive there is no inconsistency in speaking of compound atoms, and that my meaning cannot be misunderstood."

It must be obvious on considering Dalton's theory of atoms that if this is really the constitution of a gas it must also be the constitution of liquids and solids from which gases are formed or to which they may be condensed. Hence any combination ensuing between two elements implies the union of a definite integral number of atoms, *e.g.*, one atom of one kind with one or more atoms of another kind, or in more complex cases two or more atoms of one kind with two or more atoms of another kind. The consequence is that the atoms constituting each element being all alike in size and weight combination can only occur in definite proportions by weight and in the case of gases indefinite proportions by measure. The Atomic Theory was not explicitly given to the world till 1808, when the first volume of Dalton's *New System of Chemical Philosophy* was published; the subject had, however, been presented in courses of lectures given by Dalton in Edinburgh and Glasgow in the previous year. In chemistry it has long been the wholesome practice to submit every idea to the test of experiment and the fundamental law of chemical combination, that is, the law of combination in definite proportions could not well escape notice. As already stated, the combination of hydrogen and oxygen had been shown by Cavendish to occur invariably in the proportion of two measures of the former to one measure of the latter, and before the end of the eighteenth century the French chemist Proust had proved the constant composition of natural and of artificial carbonates of copper and some other compounds. This was merely a question of fact apart from any theory. Dalton, examining the analyses of nitrous oxide and of nitric oxide made by Davy, came to the conclusion that the former consists of 2 atoms of nitrogen to one atom of oxygen, while the latter is composed of one atom of nitrogen with one atom of oxygen, and here was an admission not only of definite composition in these two cases but an example of multiple proportions which as a rule or general principle Dalton established by further experiments in 1804 in the analysis of the gas from ponds (marsh gas) and olefiant gas. And so his conviction grew of the truth of the atomic doctrine, which he then proceeded to apply to chemical elements and compounds in general.

Of course Dalton occupied himself in estimating from the analysis of various compounds the *relative* weights of his atoms. Thus, for example, if 100 parts of water had been found to consist of $85\frac{2}{3}$ parts of oxygen and $14\frac{1}{3}$ parts of hydrogen or in the proportion of 6 of oxygen to 1 part of hydrogen, and on the assumption that when only one compound of any two elements is known to exist that compound is formed by the union of one atom of each, Dalton's atomic weight for oxygen was represented as 6 when that of hydrogen was taken as 1. Dalton's methods of experiment were rough and his experimental skill was of a comparatively humble order. Hence the numbers which he inserted in his tables of atomic weights were in most cases very far from the numbers which are derived from the more exact experiments of his great contemporary, the Swedish professor Berzelius, and other later enquirers into the exact values to be assigned to these ratios. Dalton's views were not immediately accepted by all the leading chemists of his day, though in a very few years the Atomic Theory was employed by all. Davy, for one, was rather slow to admit the conception of atoms, but it must be remembered that when Dalton first made his acquaintance in December, 1803, on the occasion of his visit to London to lecture at the Royal Institution, Davy was a very young man, being only about twenty-five years of age, while Dalton was about twelve years his senior. In a letter dated January 10, 1804, Dalton speaks of Davy as "a very agreeable and intelligent young man and we have interesting conversations in an evening; the principal failing in his character as a philosopher is that he does not smoke." Dr. Thomas Thomson, of Glasgow, was the chief exponent of the Daltonian doctrine, and it seems certain that the acceptance of the principle of the Atomic Theory by the chemical world was chiefly due to his expositions and writings, assisted by new observations contributed by Dr. Wollaston and other chemists. The discovery by Gay-Lussac of the simple law of combination among gases which was published soon after the issue of Dalton's *New Chemical Philosophy* afforded another very strong argument in favour of the atomic doctrine. Gay-Lassac's law of volumes states that under the same conditions of temperature and pressure gases combine together in simple proportions by measure, that is to say, one volume unites with one volume or two or three volumes of another gas, or two volumes with three, etc. Strange to say, these facts did not commend themselves to Dalton's mind and to the end of his life he seems to have been reluctant to accept the law of volumes.

It is unnecessary in this place to describe Dalton's experiments or investigations in other directions. His views on chemical subjects have long since been superseded by the results of subsequent work, and their importance was never to be compared with that of the great fundamental principles which he succeeded in establishing. But his personal characteristics require more than a few words in passing. Dalton was a man of middle stature and vigorous muscular frame. He always dressed in Quaker habit, knee-breeches, grey stockings and buckled shoes, a white neckcloth, and he usually carried a handsome gold-tipped walking cane. As to his countenance, there seems to have been a general opinion among those who were among his familiar acquaintances, and best qualified to judge, that his features closely resembled those of Newton; and a cast of Newton's head having been placed beside that of Dalton after his death it was remarked that the likeness observable during life became then more strongly marked. The life-sized marble statue by Chantrey which stands in the Manchester Town Hall represents him, it was considered, faithfully, in the prime of life. It is also said that the members of the British Association who assembled at Cambridge in 1833 were impressed by the likeness of Dalton, then living, to Roubiliac's statue of Newton in Trinity College Chapel. He preserved through life the quiet undemonstrative manners and habits which may be considered harmonious with the character of members of the body of Friends

He lived the life of a philosopher, with few wishes and little ambition, content with the most modest remuneration for his professional services either as a chemist or as a teacher. After leaving the College in Manchester and for a year or two being unsettled, he took up his abode in 1805 in the house of a friend, the Rev. William Johns, and there he remained for five and twenty years, when Mr. Johns left, and Dalton took a house for himself.

Miss Johns left an account of his life with her family from which the following extracts will sufficiently show his habits of life and something of his character:

"The Doctor's habits of life were so uniform and unvaried as to be soon related. On Sundays he always dressed himself with the most scrupulous attention to neatness, attended public worship twice, except when indisposed, or on very particular occasions, which, however, the writer does not remember to have occurred a dozen times in all; dined, during *his* life, with his friend Mr. Thomas Hoyle, the printer of Mayfield, and returning home to tea, spent the evening in his philosophical pursuits. . . . With respect to his religious principles, I should be disposed to think that he had never made theology, properly so-called, a study; he certainly never mentioned having done so, but his reverence for the great Author of all things was deep and sincere; as also for the Scriptures in which His revealed will is expressed. When the occasion called for it I have heard him express his sense of the duty and propriety of the religious observance of Sunday and also his serious disapprobation of its violation. . . . His week-days every day and all day long were spent in his laboratory, with the exception of Thursday afternoons, when he accompanied a party of friends about three miles into the country to bowl and entered into the amusement with a zest infinitely amusing to all who were present. . . . After supper we all sat together and generally had a nice chat, for which the labours of the day had excellently prepared us. The doctor took little part in the conversation, though he showed that he listened by frequently smiling and now and then uttering some dry, laconic witticism in reference to what was passing. He and my father smoked their pipes unremittingly. . . . Not unfrequently we have been favoured with the company of some of the most distinguished philosophers of Europe. . . . Among them I particularly remember M. Biot, as the most interesting person possible."

This was the general course of Dalton's life to the end, and his foreign scientific visitors might well be surprised to find the man who had achieved a world-wide reputation occupying a house in an obscure street and engaged in teaching arithmetic to young boys and girls. Notwithstanding his fame, it was not till 1822 that Dalton became a Fellow of the Royal Society, though some years earlier he had been elected a corresponding member of the French Academy.

In 1822 he visited Paris and was received by the French men of science with the utmost cordiality. He attended a meeting of the Academy and visited some of the laboratories, including those of Thénard, Gay-Lussac and Ampère, where he saw experiments which had been specially prepared for him. The following is the interesting account given by Mr. Dockray, one of his companions, of his reception by the philosophers at Arcueil:

"At four in the afternoon by a coach with Dalton to Arcueil, Laplace's country seat, to dine. On alighting we were conducted through a suite of rooms, where in succession, dinner, dessert and coffee-tables were set out; and onwards through a large hall upon a terrace, commanding an extent of gardens and pleasure grounds. It is in these grounds that are still remaining the principal Roman works near Paris, the vestiges of Julian's residence as Governor of Gaul. Avenues, parterres, and lawns, terraces, and broad gravel walks in long vistas of distance are bounded by woods and by higher grounds. As yet we had seen no one, when part of the company came in view at a distance—a gentleman of advanced years and two young men. Was it possible not to think of the groves of the Academy and the borders of the Ilyssus? We approached this group, when the elderly gentleman took off his hat and advanced to give his hand to Dalton. It was Berthollet. The two younger were Laplace's son and the astronomer royal, Arago. Climbing some steps upon a long avenue, we saw at a distance Laplace walking uncovered, with Madame Biot on his arm and Biot, Fourier and Courtois, father of the Marchioness Laplace. At the front of the house this lady and her granddaughter met. At dinner Dalton on the right hand of Madame Laplace and Berthollet on her left, etc.

Conversation on the zodiac of Denderah and Egypt (Berthollet and Fourier having been in Egypt with Napoleon), the different eras of Egyptian sculpture, the fact that so little at Rome—of public buildings—is earlier than Augustus, etc. After dinner again abroad in the beautiful gardens and along the reservoir and aqueduct of Julian. These ancient works, after falling very much into decay, were restored by Mary de Medicis. Dalton, walking with Laplace on one side and Berthollet on the other, I shall never forget."

Dalton also made the acquaintance of Cuvier, and was much struck by his interesting and attractive daughter Clémentine, whose memory he cherished ever after and whose early death he deeply lamented.

The one physical defect from which John Dalton suffered was discovered by himself. The story goes that, returning on some occasion from Kendal, he carried home with him for his mother a pair of silk stockings which he had seen in a shop-window in Kendal. "Thou hast brought me a grand pair of hose, John, but what made thee choose such a colour ? I can never go to meeting in them. Why, they are as red as a cherry." Neither he nor his brother Jonathan could perceive anything about them but a sober drab well suited for a Quaker habit. This took place when he was about twenty-six years of age, and it seems remarkable that he and his brother should have lived so long without discovering this peculiarity in their vision. Dalton read a paper on the subject of colour vision before the Philosophical Society of Manchester in 1794, and the defect having been found to be more commonly prevalent than had been supposed, many investigations have taken place since that day. It is obvious that the confusion of red with green would be seriously objectionable in a seaman or signalman on a railway. The Board of Trade has instituted a series of tests for persons entering such employments, and a proper examination is or ought to be compulsory upon all.

It may be imagined that Dalton's defect of vision and his Quaker habits and opinions were often sources of embarrassment to his friends if not to himself. Thus in 1832 he received at Oxford, together with Faraday, Robert Brown and Sir David Brewster, the honorary degree of D.C.L., which required him to wear for the occasion the doctor's scarlet robe. However,

"You call it scarlet," said Dalton; "to me its colour is that of nature, the colour of those green leaves." The same gown came in useful when he was presented to King William IV.; Court dress would have involved wearing a sword, which he refused to do, but the doctor's robe concealed all that in the eyes of the Court would have been regarded as irregular.

In 1826 the Royal Society awarded to Dalton the first of the Royal Medals given by the king, and this afforded to Sir Humphry Davy as President the opportunity of testifying to the high position held by the medallist in the esteem of the world of science. "Mr. Dalton's permanent reputation," he said, "will rest upon his having discovered a simple principle universally applicable to the facts of chemistry, in fixing the proportions in which bodies combine and thus laying the foundation for future labours respecting the sublime and transcendental parts of the science of corpuscular motion. His merits in this respect resemble those of Kepler in astronomy."

In 1833 the Government of Lord Grey granted a pension of £150 a year to Dr. Dalton and in 1836 this was increased to £300. The announcement of the first grant was communicated to Dalton by Professor Sedgwick in his address as President of the British Association delivered in the Senate House at Cambridge. In 1834 Dalton sat to Chantrey, the sculptor, for the statue destined to stand in the Manchester Town Hall.

In the same year his brother Jonathan died, and the manner of his decline from paralysis seemed to foreshadow the end of his distinguished brother, for both survived the first seizure several years. The first attack, in the case of John Dalton, happened in April, 1837. From this he partly recovered, but on July 27, 1844, the malady recurred and was followed by a peaceful death a few hours later.

Dalton's career illustrates in an interesting way the fact that independence of character may lead a powerful mind to conclusions of the greatest importance, notwithstanding deficiencies resulting from imperfect education. John Dalton conceived and showed the application of the Atomic Theory to chemical facts. That his own estimations of the atomic ratios of the chief elements were far from exact seemed to have little weight with him, neither does such a fact diminish the merit of the conception. All subsequent developments of the science of Chemistry have been dependent on the idea, and no one now pretends to dispute the validity of the Atomic Theory in physics and chemistry. The discovery that atoms are complex structures capable of disintegration nowise affects the recognition of the atom as the elementary unit.

As a man and a philosopher Dalton presented a remarkable contrast to his great contemporary Davy; they could have had little sympathy with each other except in their devotion to science and the expansion of human knowledge. The one simple in dress and manner, often uncouth in speech and unskilful in experiment, the other courted by society, an eloquent and attractive speaker, a brilliant lecturer, a dexterous experimenter and withal filled with poetic enthusiasm. Dalton seems to have modestly recognised the limitation of his own powers, attributing his success largely to "perseverance." Davy, on the other hand, encouraged by early triumphs, was apparently always ready to accept the flattering estimate by contemporaries of the importance of his successive discoveries.

Joseph Louis Gay-Lusac.
Professor in the University of Paris.

[Face p. 119]

CHAPTER IX

GAY-LUSSAC

THE simple law of combination among gases already referred to was discovered by the French chemist, GAY-LUSSAC. The course of his scientific work corresponded in its early stages nearly to that of Dalton, inasmuch as he began with meteorological observations, though of a much more showy and adventurous character.

Joseph Louis Gay-Lussac was born on September 6, 1778, at St. Léonard, a small town in the department of Vienne near the borders of Auvergne, where his grandfather had practised medicine and his father held office as Procureur du Roi. The revolution of 1789, which broke out while he was yet a child, obliged his parents to keep him at home during all the years when he would have been receiving in better times a classical education. It was only in 1795, when he was more than sixteen years of age, that it became possible to go to Paris to begin his studies and to prepare for the examinations for admission to the École Polytechnique. Unfortunately a great scarcity having arisen, M. Lavouret, the head of the establishment he had entered, found himself obliged to dismiss all his boarders. He was, however, transferred to the care of M. Censier, whose establishment for some time at Nanterre and afterwards at Passy enjoyed some advantages of which the pensions in the capital were at that time deprived. But difficulties in obtaining food soon reached the Censier's family, and the boarders were dispersed. Gay-Lussac had happily attracted notice alike by his talents and his amiable qualities, and he was retained almost more in the character of a son than of a pupil.

As illustrating the hardship to which the family was at this time exposed, it appears that Madame Censier, who kept two cows in her garden, carried the milk every night into Paris in order to sell it, but as the roads were unsafe she was accompanied by young Gay-Lussac armed with a big sabre. On the return journey, which was accomplished in daylight, the boy occupied himself at the bottom of the cart in studying geometry and algebra in preparation for the approaching examinations of the Polytechnic. He found himself therefore ready to enter the École Polytechnique at the end of 1797, and in three years' time he obtained a post in the service of the "ponts et chaussées" into which the best students then flocked.

Berthollet was at this time professor of chemistry in the École Polytechnique. He noticed the activity and intelligence displayed by his pupil, and Gay-Lussac became his demonstrator. Gay-Lussac also assisted Fourcroy. Soon afterwards the professor carried him off to his private laboratory at Arcueil, where, surrounded by all necessary appliances, he was occupied in working at his ideas concerning chemical statics, stimulated by the daily visits of his friend Laplace, whose residence a little later adjoined his own. It was under the influence of these two men that Gay-Lussac commenced his career.

He was thus led to devote himself to that field of research common to physics and chemistry which had already been partly explored by Dalton. Accordingly in 1802 he published in the *Annales de Chimie* his first essay on the dilatation of gases and vapours; then immediately following this work he engaged in a series of researches on the improvement of thermometers and barometers, on the tension of vapours and their diffusion into gases, the determination of vapour densities, the laws of evaporation, hygrometry, and the measurement of capillary effects. An opportunity then offered for the utilisation of the knowledge he had obtained of these subjects. He was commissioned, together with his friend Biot, to make a balloon ascent in order to find out if it was true, as supposed, that magnetic forces were no longer exerted at a distance from the surface of the earth. They established the fact that, on the contrary, magnetic effects continued undiminished up to an elevation of 4,000 metres. The balloon employed was too small to be capable of carrying the two men together to a greater elevation. Gay-Lussac thereon made a second ascent *alone*, and reached an elevation of 7,000 metres, the highest ever attained up to that time. He confirmed the previous observation of the persistence of magnetic forces, and he brought back samples of the air of these elevated regions, which when submitted to analysis were found to have the same composition as the air at the surface of the earth. He also made a series of valuable observations which established the gradual reduction of pressure temperature and moisture during his ascent. There can be no doubt that the results of these observations represented a great advance on anything previously known in respect to the atmosphere. It was, however, necessary to determine more exactly the principal properties of gases. Thus the co-efficient of expansion of gases as determined by Gay-Lussac was more exact than that published by Dalton, but was still far from the truth. Like Dalton, he also adopted the erroneous idea that the expansibility of all gases was the same, and he further supposed the co-efficient to be constant, whereas it varies with pressure and temperature, though admittedly only to a small extent. Supposing a mass of gas to have unit volume at 0° C. under mean atmospheric pressure and it is then heated to 100° C. under the same pressure, the volume becomes, according to Dalton, 1·3912, while according to Gay-Lussac it is 1·3750. Later experiments by Regnault and others have shown that both these figures are too high. Possibly Gay-Lussac himself realised that with the knowledge and the appliances of his day such researches could be carried successfully no further.

In 1805, by the friendly intervention of Berthollet, Gay-Lussac was released from duty for the purpose of accompanying Alexander von Humboldt in his travels into Italy and Germany. He took with him an outfit of instruments for magnetic and meteorological observations. Starting from Paris on March 12, 1805, and passing over Mont Cenis and by way of Geneva, the travellers reached Rome early in July. The society of his distinguished companion, not long returned from his famous travels in the American continent, was of course a great advantage to the young chemist. Moreover, he then for the first time beheld the wonders of Alpine scenery. In Rome also, in the house of Wilhelm von Humbolt, who was Prussian Minister, he enjoyed the society of artists and savants, and even had the use of a chemical laboratory where he discovered the presence of fluorides as well as phosphate in the bones of fish.

Soon after this von Humboldt and Gay-Lussac, in company

with von Buch, then a young geologist, proceeded to Naples. Vesuvius, which had been for some time in a quiescent state, suddenly started into activity, and Gay-Lussac had the "happiness," according to the expression of one of his companions, to experience one of the most tremendous earthquakes ever known at Naples. He ascended Vesuvius six times for the purpose of making observations. On September 17th the three friends left Naples and arrived at Florence on the 22nd, and after a few days went on to Bologna. At Milan they visited Volta, though they had some difficulty in finding him, living as he did in obscurity. Crossing the St. Gothard, where the pleasure anticipated was destroyed by the occurrence of fog, the travellers reached Berlin in the middle of November. Gay-Lussac was retained as the guest of von Humboldt all the winter, and returned to Paris in the spring of 1806, and soon afterwards was elected a member of the Institute.

In 1807 Berthollet formed a special scientific society composed of a very small number of members which was known as the *Société d'Arcueil*, from the name of the commune near Paris in which was situated his country residence. Gay-Lussac was among the first members, and the first volume of *Mémoires* issued by the society commences with a record by Gay-Lussac of the results of all the magnetic observations made in concert with Humboldt during their recent travels in France, Italy and Germany. The second volume of *Mémoires* contains a paper on the combination of gases among themselves. It had been known from Priestley's time that hydrogen combines with about half its bulk of oxygen, and Cavendish's experiments rendered it certain that the true ratio between the volumes of the combining gases in this case was exactly 2 to 1, at any rate that the observed difference did not exceed the amount fairly attributable to errors of experiment. Gay-Lussac observed that correspondingly simple relations exist whenever two gases are found to combine together chemically. Thus Gay-Lussac proved that two volumes of carbonic acid are formed by the union of two volumes of carbonic oxide and one volume of oxygen, also that hydrogen and chlorine unite in equal volumes and hydrochloric acid and ammonia also in equal volumes. He also showed that two volumes of ammonia are composed of one volume of nitrogen and three volumes of hydrogen, and in many similar cases a like simple proportion was to be observed. Pursuing the idea, he presented it as a definite proposition in 1808, and it has since been established on a firm experimental basis as one of the fundamental propositions generally accepted in chemistry.

About the same time Gay Lussac, associated with Thénard, devised a method by which the metals potassium and sodium, recently discovered by Davy, could be obtained in larger quantity than was then possible by the electrolytic method of Davy. Their first process consisted in the decomposition of melted potash or soda by means of incandescent iron. For fifty years or more the chemical process by which potassium and sodium, were distilled from their carbonates mixed with carbon at a high temperature was universally employed in the commercial production of these metals, but it is interesting to note that the invention of the dynamo and the development of electrolytic methods applicable on an industrial scale has led to a return to the simple process discovered by Davy. The process of Gay-Lussac and Thénard, however, was at that time more productive than Davy's and thus the remarkable properties of the metals potassium and sodium could be more conveniently studied. One of the first results obtained by Gay-Lussac and Thénard was the isolation of the basis of boracic acid, which they denominated *bore*, or, as it is now called, boron.

During 1808 and for some time there was active rivalry between the French chemists and their English contemporary alike in experimental research and in theory. The nature of potassium and sodium, for example, was the subject of difference of opinion in which ultimately the French chemists, who supposed these metals to be hydrides of the alkaline oxides, admitted that they were wrong. Similarly the French chemists, in adhering to Lavoisier's view of the nature of oxygen as the only acidifying principle, and neglecting Davy's experiments which proved that chlorine (then called *oxymuriatic* acid) contains no oxygen but is entitled to be ranked among the elementary substances, found themselves again obliged to change their theories. The establishment of chlorine among the elements proved an important guide in the study of the interesting new substance discovered later by Courtois, a French manufacturer, in the mother liquors left after the extraction of soda from seaweed. This substance was brought under the notice of Gay-Lussac, and on December 6th he read a short paper to the Institute in which he described the chief properties of the new substance to which he gave the name *iode* (English iodine). A fortnight later he described iodic acid.

About this time, as already related, Sir Humphry Davy was in Paris, and having become possessed of a small quantity of the new substance through the agency of Ampère, he set to work to examine it, and at the request of his friends among the French chemists, he read a note to the Institute in the interval between Gay-Lussac's two communications, in which he gave the result of his observations, including the recognition of the fact that the acid formed by the new substance was not muriatic acid, as had been previously supposed. The analogy recognised between iodine and chlorine afforded much assistance, and the chemical history and characteristics of iodine were fully established in the following year, by the masterly work of Gay-Lussac. Here was also an opportunity for the application of his law of volumes, whereby it was possible to calculate the density of the vapour of iodine which was not then known, and this bold speculation was subsequently confirmed. A year later, in 1815, Gay-Lussac in his researches on prussic acid discovered cyanogen, the first known compound radical. This substance, in forming compounds with metals by the interaction of prussic acid and metallic oxides, reveals its analogy with chlorine and illustrates farther the inadmissibility of Lavoisier's views about acidifying principles.

Gay-Lussac became in 1809 professor of chemistry in the *École Polytechnique* and at the same time professor of physics at the Sorbonne. He subsequently held many public offices, and in 1832 he relinquished the chair at the Sorbonne to take the professorship of chemistry at the *Jardin des Plantes*. In connection with his official work, for example, in his later years as assayer to the Mint, he was led to devote much time to the practical applications of science. For example, in the manufacture of sulphuric acid by the Chamber process he introduced the use of the tower which has always borne his name, for the purpose of absorbing and utilising the oxides of nitrogen which otherwise pass into the air and are lost. He may also be regarded as the originator of volumetric analysis, for in lieu of the cupellation process universally employed up to that time in the assay of silver, he introduced the wet method by which greater accuracy and saving of much time was secured.

Gay-Lussac's great talents were devoted uninterruptedly to

the cause of science alike in his earlier years, when he was occupied with the extension of knowledge, and in his later professional life when the duties of the several offices he held required undivided attention.

We have already seen (p. 95) the high opinion of him expressed by Davy. As a teacher he employed a clear and simple language always adapted to the subject under discussion. He was acquainted with the German and English languages, and took care to read assiduously the publications relating to physics and chemistry in foreign countries. Few chemists who pursue industriously experimental research escape without accidents more or less serious, and Gay-Lussac was the victim of several mishaps in the laboratory, of which one explosion in which potassium was concerned nearly cost him his sight. Those of his friends who were familiar with the cool reserve of his ordinary manner found it difficult to believe that in his boyhood he had been turbulent and adventurous. He used to tell stories of escapades in which he was the leader, and of which the robbing of an orchard was generally the object.

The story of his marriage is full of romance. At the beginning of the Revolution in 1789 there lived at Auxerre a musician attached to the college in the town. On the suppression of these establishments in 1791 it became necessary to educate his three daughters with a view to gaining their living as teachers. But the eldest, Josephine,[1] in view of the family difficulties, preferred to take a situation in a linen-draper's in Paris, and there Gay-Lussac made her acquaintance. The young lady behind the counter he noticed to be reading attentively a small book which on enquiry turned out to be a treatise on chemistry. Naturally the interest in such a subject displayed by a girl of seventeen excited his curiosity, and something more, for his visits to the shop became more and more frequent and in the end the young lady accepted his offer of marriage. Gay-Lussac then placed her in a school in order to finish her interrupted education, and in no long time, namely in 1808, she became his wife. The tender sympathy subsisting between Gay-Lussac and his wife during forty years controlled so completely their actions and even their habits as to extend even to their handwriting, and in the

---

[1] The family name is not given in Arngo's biographic notice of Gay-Lussac.

end it was impossible to distinguish the manuscript of a memoir copied by Madame from the original as it proceeded from the hand of her famous husband.

Gay-Lussac was created a peer of France by King Louis Philippe. Berthollet had been a senator during the Empire and a peer of France at the Restoration, and dying in 1822 he had left as a legacy to Gay-Lussac the sword which was part of his costume as a peer. This testamentary disposition excited surprise at the time, but it arose clearly from the feeling on the part of the venerable academician that Gay-Lussac, as the most distinguished among the French chemists of his day, was his legitimate successor.

Prejudices among the old aristocracy prevailed for a time, and Gay-Lussac had to wait for his peerage till 1839. France has always regarded with just pride, and posterity will look back with respect and admiration to, the career of this distinguished man. He died on May 9, 1850, in his seventy-second year.

PROUST

[Face p. 127]

CHAPTER X

PROUST

THE name of PROUST is not so familiar to the modern chemical student as his great services in helping to lay the foundations of fact necessary to the establishment of chemistry as a branch of science, when recognised, would lead us to expert. For it was by his experiments that the fundamental principle of constant proportions in chemical combination was brought to such a position as to be generally recognised. It appears now to be obvious that the assumption of a constant composition in many compounds was indispensable in the various analytical processes practised long before the end of the eighteenth century. It was known, for example, that there was a constant relation between the quantity of an acid neutralised by a base, and the quantity of the base required for the process of neutralisation. Such facts had been published by two German chemists, Richter and Wenzel, long before the end of the eighteenth century. Their results, however, had been obscured for a long time by their retention of the phlogistic doctrine, which, as already mentioned, enveloped in fog so many writings of the time.

Joseph Louis Proust was born at Angers in 1755, the son of a *pharmacien* of that town, and in his youth was destined to follow his father's profession. He therefore began his studies at home, and in due time was sent to Paris to complete his training under the direction of M. Clerambourg, an apothecary of repute at that period. He devoted himself with enthusiasm to the study of chemistry and the practice of his art, thus recalling the early career of the illustrious Swedish chemist Scheele, only a few years senior to Proust. The position of chief *pharmacien* at the hospital of La Salpétrière having fallen vacant, Proust was appointed to the post and thus gained the means not only of an honourable livelihood but of pursuing the study of chemistry, which had for him an irresistible attraction. In this direction he had the advantage of encouragement and advice from Rouelle, then a famous teacher, in whose class he had been a diligent pupil. While at the Salpêtrière Proust published notes and memoirs on a diversity of subjects, in which he displayed that independence of spirit which characterised all his scientific work in later years. He also taught chemistry both at the Palais Royal and in an establishment founded by the unfortunate Pilatre de Rosier. He accompanied this physicist in his first balloon ascent, which took place at Versailles in June, 1784, in presence of the Court and of the King of Sweden. After a few preliminary difficulties the balloon, of the Montgolfier type, distended by means of heated air, ascended to considerable height out of sight of the anxious spectators. The balloon descended after a journey of upwards of an hour at a distance of some forty miles from the point of departure, and without injury, but Proust, though finding himself safe, had calculated the danger and refused to ascend again. Pilatre some time later lost his life in a second ascent in which two balloons were employed, the upper one filled with hydrogen, the lower with heated air. As Proust had foreseen, the hydrogen caught fire, and Pilatre and his companion fell and were killed.

The Spanish Government, observing the progress in the physical sciences then conspicuous in France, and stimulated thereby in the hope of benefit from its application to the purposes of industry, offered Proust a post at the School of Artillery at Segovia. This post he accepted, and soon justified the reputation which he had acquired in his own country. The Government determined to institute at Madrid a central school of chemistry, and Proust was appointed Professor. At the same time the establishment was furnished with instruments and appliances on a magnificent scale; almost all the vessels, even those in common use, were of platinum. In this laboratory and in the laboratory at Segovia Proust's principal researches were carried out, including many the purpose of which was the immediate application of the results to practical purposes. Thus he analysed many Spanish minerals and other natural products, beside enquiring into questions connected with food, including gelatine, Iceland moss, and grape-sugar. Unhappily in the war then proceeding in the Peninsula the French occupied Madrid and he was at one stroke deprived of his post and his splendidly equipped laboratory, beside the collection of minerals and chemical products he had accumulated with so much labour. Napoleon, desiring to encourage in France the manufacture of grape-sugar, commissioned Proust to set up at the cost of the Government a factory for its production together with the offer of a liberal fee, but in consideration of his age and the state of his health he felt obliged to decline the offer and he retired to Craon (Mayenne), where he lived on his small patrimony. In 1816 Proust, though not a resident in Paris, was by a special decree nominated member of the Academy of Sciences on the death of Guyton de Morveau. A small annual payment being associated with the title of Academician, added to a pension of 1,000 francs accorded to him by King Louis XVIII., contributed to render easier the circumstances of his later years. After his election as Academician Proust returned to his native town, Angers, where he continued to pursue his chemical researches to the end of his life. He died on July 5, 1826.

Proust's most important work, and that on which is founded his title to be enrolled among the founders of modern chemistry, consists in a series of researches analytical and synthetical, by which he established the fact that chemical combination occurs only in definite proportions by weight, and that every chemical compound is definite in its nature, the ratio between its constituent elements being invariable.

At the end of the eighteenth century the man who occupied the position of perhaps greatest authority in chemistry was Claude Louis Berthollet (1748-1822). After the revolutionary period he had become a professor in the Normal School and in the Polytechnic, and took part in Buonaparte's expeditions to Italy and to Egypt. Later he received the highest honours as member of the French Academy of Sciences, and under Louis XVIII. became a peer of France. He died at Arcueil, near Paris, in 1822. He rendered considerable service to industrial chemistry by the application of chlorine in the form of chlorinated lime to the purposes of the bleacher and dyer, but his researches and theories in connection with the laws of chemical combination attracted the greatest interest. They were embodied in a famous work under the title *Essai de Statique Chimique*. According to the view therein maintained, the force of chemical affinity, like that of gravity, must be proportional to the masses of the acting materials, and the composition of any given chemical compound might therefore vary according to the circumstances attending its formation. Soon after the beginning of the nineteenth century the number of cases in which elements were found to unite together in definite proportions was gradually increasing, and the observation of combination in multiple proportions by John Dalton rendered this doctrine more and more associated with difficulty. At the same time Proust, then professor in Madrid, definitely took up the challenge implied in Berthollet's work. Before the end of the eighteenth century Proust had proved that carbonate of copper, whether in the form of the natural mineral or prepared artificially by precipitation, had uniformly the same composition. He showed a little later that the oxides of tin and the sulphides of iron were also definite in composition, and further, that when more than one compound is formed by any two elements the proportions increase by leaps and not gradually. He proved also that Berthollet's analyses, by which he supposed that he had proved that metals can unite with gradually increasing quantities of oxygen, were incorrect, as many of the substances employed were not definite in their nature, but consisted of mixtures. It is remarkable that, having established so many facts, he did not take the one further step which would have led to the discovery of the law of multiple proportions. Up to this period composition had generally been recorded in percentages, but when one element is taken as a fixed quantity it is found that the other varies according to a simple rule. Thus in 100 parts of red copper oxide there are 88·8 parts of copper with 11·2 parts of oxygen, while in the black oxide 79·87 parts of copper are associated with 20·13 parts of oxygen. In the red oxide, therefore, 1 part of copper is combined with ·126 part of oxygen, and in the black oxide 1 part of copper is united with ·252 part of oxygen. It is evident, therefore, that in the black oxide the amount of oxygen present is exactly twice as great as the amount of oxygen in the red oxide for every unit weight of copper. This discovery was reserved for John Dalton, as already explained, and the principle of definite and multiple combination was accounted for by his Atomic Theory.

[*Face p.* 131]

CHAPTER XI

BERZELIUS

It has already been mentioned that the Atomic Theory introduced by Dalton was accepted by the chemical world only gradually and almost reluctantly. Perhaps this was partly attributable to the lack of exact experimental evidence, for Dalton's own work was crude and the combining weights of the elements assumed by him from the results of his own experiments were very far from the values established later. The general recognition of the doctrine was, in fact, chiefly due to the laborious investigations of the great Swedish chemist Berzelius, by whom the *quantitative* as distinguished from the *qualitative* era in chemistry may be said to have been inaugurated.

Jöns Jacob Berzelius was born in 1779, and was thus about thirteen years younger than Dalton.

An autobiography has come down to us written originally in Swedish, but translated into German by Fraülein Emilie Wöhler, daughter of the famous chemist of that name, who had been a pupil of Berzelius. From the preface to this work we learn that the Statutes of the Royal Academy of Sciences of Stockholm require every member of the Academy on election to prepare a biography of himself and to bring it up to date every ten years. The author explains that he fulfilled this duty to the end of the year 1840, and that what he has composed is obviously much in excess of the requirements of the Academy. This is very fortunate, for the autobiography not only corrects some misstatements proceeding from other sources, but contains many details of great biographical interest. The whole occupies about 112 pages, octavo, in No. vii. of Kahlbaum's *Monographs from the History of Chemistry*, dated 1903. It has never appeared in English, but the following pages contain a version, necessarily much condensed, supplying the most important facts of which an account is given in the original narrative.

Jacob Berzelius was the son of Samuel Berzelius, principal of the school in Linköping, in East Gothland, and was born on August 20, 1779, at Väfversunda, where his parents were staying during holidays in his mother's former home. His father died of consumption in 1783, and his mother then returned to Väfversunda, where she remained several years, till she married the German pastor at Nörkoping, Anders Ekmarck. Ekmarck was a man of exemplary character and was a good father to his stepchildren. He had two sons and three daughters by a former marriage, and they all lived happily together till Berzelius's mother died, and one of her sisters, Flora Sjösteen, undertook the care of the house and the children. Ekmarck then removed to Ekeby, where the education of the children was carried on partly by a tutor named Haglund, and partly by Ekmarck himself, who seems to have encouraged them in the study of nature, especially of plants. He sometimes said to his stepson, "Jacob, thou hast talent enough to follow the footsteps of Linnæus; fear God always and thou will do so all the more surely." These words of praise, however, had the effect of leading the boy to fancy that something great was in store for him without much exertion on his part.

In 1790 the aunt who was keeping his step-father's house married, and Ekmarck entered a third time into matrimony, this time with a widow who already had a daughter. Under these circumstances Berzelius and his sister were taken in by their uncle Magnus Sjösteen, who had seven children of his own. This change was an unhappy one, for although the uncle made no difference in the treatment of the nephew and his own children, the latter were not only unruly but unfriendly to their cousin.

In 1793 Berzelius was sent with the rest to the Linköping Gymnasium, where he made rapid progress, but the next year, being very unhappy, he obtained a post as tutor in the house of a farmer at Norköping. Of the two sons of the farmer one was fifteen, about the same age as Berzelius, the other a year older, but as they only wanted to learn the catechism, their young teacher had plenty of time. He found his former beloved tutor, Haglund, who now called himself Hagert, at a neighbouring village, and joined him in collecting plants and especially insects. Hitherto Berzelius had looked forward to becoming a clergyman, like his father, his grandfather, and his great-grandfather, but his studies in natural history now led him to think of the medical profession. During the greater part of the year he spent in the house of the farmer he worked on the farm. In winter he chopped wood and brought it to the house. This, as he says, was a mean business, because the wood, which could not be bought in this treeless district, was taken unlawfully from the common stock. His own room would not have been heated in the winter but for the fact that it was necessary to keep the frost from the potatoes which were stored there.

In his spare time he worked at a translation of Virgil's *Æneid*, and this was useful as evidence on his return to the Gymnasium that his studies had not been neglected. His health benefited greatly by the air and exercise on the farm, and he returned to school in February, 1795, with unexpected development of his previously slender and unhealthy frame. On leaving the farm he received four Reichstalers and a pair of woollen stockings, of which he "stood greatly in need."

At the Linköping Gymnasium there was at this time a new and energetic teacher of natural history, C. F. Hornstedt, under whose influence Berzelius made great progress in knowledge of nature and in power of observation. He soon belonged, as he says, "to wood and field rather than to town and school." Fortunately only the morning and evening hours were generally productive of specimens, otherwise the fascination of these pursuits, especially with the gun, would have cost him many hours of school instruction. The somewhat careless use of the gun and the mere fact of carrying it contrary to rules of the school, brought down on him the displeasure of the authorities, who looked unfavourably on his natural history pursuits and persecuted the teacher by whom he had been encouraged.

In the winter of 1795 Berzelius joined some of his comrades in forming a literary circle in which their own compositions in prose or verse were read. They were visited by Bishop Lindblom and from him received approval and encouragement, which were useful to Berzelius as an Abiturient on joining the University. The rector had stated on his certificate that he was "a young man of good abilities, but bad habits and doubtful ambition, and charged him with having wasted sixty-three hours of the semester. Here the Bishop spoke up for him and dismissed him with words of encouragement.

In September, 1796, Berzelius went to Upsala hoping with his small inheritance to carry on to the end of the spring semester. This was insufficient, but after taking a post as private tutor for a year he was recalled to the University with the offer of a scholarship. This, with the small sum remaining of his own money, was all that he could reckon on for completing his education.

In the autumn of 1798 he passed the medico-philosophical examination, *non sine laude approbatus*, though he had learned so little chemistry at that time that he was only allowed to pass on the strength of his performance in physics.

His oldest half-brother, Lars Kristofer Ekmarck, had been at the University two years, and having become interested in electricity he helped Berzelius to make experiments. The two young men obtained Girtanner's *Anfangs Gründe der anti-phlogistischen Chemie*. This was the first German text-book on the new system, and appeared in 1792. So far the two students had never seen one of the phenomena about which they had read. Berzelius accordingly applied to the professor, Johan Afzelius, for permission to work in the laboratory. What he wanted to see was how to prepare and collect a gas, and the burning of carbon, phosphorus, iron, etc., in oxygen. The professor, who had but little inclination for the new doctrines, put obstacles in his way, but ultimately left him to his own devices.

In the following semester Berzelius rented a student's apartment having a small windowless room adjoining with a fireplace, and here he carried on experiments. "One day," he says, "as I was making fuming nitric acid and noticed some gas escaping, I collected it over water in some bottles I had at hand, to find out, if I could, what this gas was. I suspected oxygen, and seldom have I had a moment of such pure and heartfelt joy as when the glowing splint placed in the gas burst into flame, and lighted up my dark laboratory. On the next working day I prepared several bottles of oxygen in which iron wires were burnt amid general gratification."

At the beginning of 1799 Berzelius began the study of anatomy, and this, being now the principal subject, required the greatest diligence. Chemistry was his recreation in the daily hours of freedom. In this direction, however, he got little encouragement. His uncle, Daniel Berzelius, an apothecary at

Jönköping, refused his request for admission into his pharmacy, and though invited to stay with his aunt, Flora Bromander, in Bromma, her husband, a man of business who had begun life as a cattle-dealer, had little sympathy with a mere student, and soon sent him off to an apothecary, Wessel, in Vadstena. He found that there was but little to learn in the pharmacy. Here, however, he made the acquaintance of an Italian glass-blower, Joshua Vaccano, whom he describes as a good-natured old fellow, and from him he learned to blow glass and to make barometers and thermometers. After a month he returned to Bromma, and by the introduction of Bromander he became assistant to Sven Hedin, physician at the Medevi Springs, and there he remained during two summer seasons. He took the opportunity to analyse the water of the Medevi and Loka Springs, the residues obtained by evaporation being carried to Upsala. This was his first attempt at analysis. Bergman's analysis of Medevi water, published so long before as 1782, was his sole guide, as he could get no help from his teachers. He desired to use the account of his analysis, which, with Ekmarck's help, he translated into Latin, as his dissertation. After encountering various difficulties arising out of professional punctiliousness, petty formality and his own poverty, Berzelius at last obtained his certificate in May, 1801.

Just at this time Volta's discovery of the electric pile became known, and Berzelius immediately constructed one with sixty pairs of copper coins and zinc plates and took it off to Medevi with the object of trying its effect on patients at the baths. He only succeeded in producing a beneficial effect in one case, that of a soldier who had lost the use of one hand. His salary came to an end in June, but Dr. Hedin procured for him a surgical post with the aid of which he proceeded to the licentiate examination in the autumn of the same year, and in May, 1902, having completed all the necessary preliminaries, he at last obtained his degree as Doctor of Medicine. It is not surprising to learn that so many disappointments, so much hard work and privation were followed by a serious illness which for two months put a stop to his activities. An illustration of the difficulties encountered by a student such as Berzelius is provided by the story of a paper he had prepared on nitrous oxide, then a comparatively new substance. Having placed it in the hands of Professor Afzelius to read, it was sent on to the College of Medicine and thence to the Academy of Sciences. Three years later he learned from the secretary that the Academy would not publish the paper because "they did not approve the new chemical nomenclature."

On his recovery from illness Berzelius proceeded to Stockholm, where he was offered a couple of rooms in the home of W. von Hisinger, who joined him in experiments on the chemical actions of the pile. A report of this work in their joint authorship appeared in *Gehlen's Journal der Chemie* early in 1803,[1] and contained an expression of views which formed the basis of the theory of chemical combination made use of by Berzelius at a later time. An attempt to give a series of scientific lectures failed for want of subscribers, and an attempt to manufacture mineral waters was equally unsuccessful commercially. Medical practice at the Springs yielded only a small income.

In conjunction with his friend Hisinger he made the discovery of the new element cerium, but appears to have been anticipated by Klaproth, who had given the name "ochroiterde" to the new earth. However, Berzelius refers to the fact that he and Hisinger had enjoyed "the triumph of having described its character more correctly than the greatest analytical chemist in Europe." In those days a distinction was made between an earth and a metallic oxide. The mineral, thus made for the first time a subject of analytical study, has had a long subsequent history from which the difficulties of the research are obvious.

In Stockholm Berzelius describes his life as somewhat solitary. His friend Werner, with whom he had been associated in the attempt to establish a mineral-water business, and in whose house he resided for some time, was never at home at meal times and lived very much in the various clubs to which he belonged and to which he was ready to introduce Berzelius. Werner was no man of business, and an attempt to establish a vinegar factory resulted in his bankruptcy, and involved Berzelius as surety for loans to Werner from the bank in a debt which it took him ten years to pay off. "I was, however," he says, "not discouraged; I reflected as usual, Kommt Zeit, kommt Rat." (Things will come right in time.)

[1] This must be a mistake for 1806, as *Gehlen's Journal* began in that year.

In 1805 he undertook the analysis of the water from the Adolfbergsbrunnen and received from the company a liberal fee which enabled him to make an excursion to Fahlun, where he made the acquaintance of Gahn, the mineralogist, and to visit the iron works and mines in Dannemora. The waters he had analysed presented novel points of interest in the presence of potassium and manganese carbonates, and the evolution of considerable quantities of nitrogen gas. He was, however, again unlucky in his offer of a paper on the subject to the Academy of Sciences, and he resolved to send no more papers to the Academy.

Having unsuccessfully contested the secretaryship of the Medical College, Berzelius was in 1806 appointed Reader in Chemistry at the Karlberg Military Academy in succession to Professor Quensul, who had died suddenly, and in January, 1807, he was appointed Professor at the School of Surgery at Stockholm, and as he was able to retain the post of Armenarzt (physician to the poor at the Springs), he had now a tolerable income. The Medical College at this time had a house of its own, and Berzelius was enabled to carry on his pharmaceutical and chemical work with the aid of a grant of money from the state for instruments and materials. He shared rooms with Dr. Pontin, his assistant, and fitted up the kitchen as a laboratory.

In 1807 Berzelius had the advantage of spending a summer holiday of some weeks in the south of Sweden and in Copenhagen, where he made the acquaintance of the famous physicist Oersted, with whom he established a friendship which lasted through life.

For some time previously he had been engaged in the preparation of a text-book of chemistry in Swedish, to take the place of the German books previously in use. This was followed by a volume on physiological chemistry, and the whole ultimately expanded into five volumes which were afterwards translated into German by Wöhler.

While working at this book his attention was attracted to Richter's researches on the reciprocal decomposition of neutral salts and the resulting neutrality of the products. This was the starting-point of the researches on chemical proportions to which the greater part of his scientific working life was devoted. His first experiments were not very successful, owing both to want of experience and lack of apparatus. There was but one platinum capsule in all Sweden, and this belonged to Hisinger. It was too heavy for his balance. Berzelius, in his autobiography, describes in detail the difficulties he encountered and how they were overcome.

His position, in the meantime, had considerably improved. The Academy of Sciences had determined to issue a journal under the title *Œconomische Annalen* to which Berzelius was invited to contribute articles, either original or translations, for which a liberal honorarium was offered. In the second year of its existence it was under his editorship. But a more important event was the establishment of the Karolinisches Medico-Chirurgisches Institut on the reorganisation of the medical school to which Berzelius was attached. This resulted in the creation of several new professorships, in consequence of which he was released from the necessity of lecturing on medicine and surgery, while retaining chemistry and pharmacy.

He was now in a position to give up the post of doctor to the poor, and so got more time for scientific work. A mishap unfortunately occurred which might have put a stop to it for ever. In the attempt to recover the metal from some fulminating gold an explosion occurred which so damaged his eyes that for more than a month he was confined to a dark room, and long after suffered in the right eye.

In 1808 Berzelius was elected an ordinary member of the Academy of Sciences, and from February to August he held the office of president.

The sudden death of the Crown Prince, Karl August, in 1810, had led to the suspicion that he had been poisoned. This was followed by an enquiry in which Berzelius was involved, and the waste of much time. He had reported that there was no evidence of poisoning, but in consequence of the unsettled state of Sweden at that time the matter remained a subject of suspicion both on the part of the populace and the authorities.

In 1811 he became member of a committee which lasted till 1818 on the production of saltpetre. This brought him a substantial addition to his income and a further increment on receiving the joint secretaryship of the Academy of Sciences. The latter involved the renunciation of his vow to send no more papers to the Academy, one of the conditions of appointment being that the results of his work should be communicated to the Academy for publication.

By this time Berzelius had many correspondents in foreign countries, among the rest Berthollet and Davy. The former sent him an invitation to Paris, and funds for the expenses of the journey were promised by the Crown Prince (Bernadotte). Before he was ready to start, however, war broke out between Sweden and France. Berthollet sent him a passport, but at the wish of the Crown Prince the projected visit to France was abandoned and an excursion to England substituted. At the end of June he landed at Harwich and arrived the same evening in London. Except Copenhagen, Berzelius had previously never visited a foreign country, and he spoke but little English.

Davy was then the most famous chemist in Europe, and the chief object of the journey to England was to make his personal acquaintance. Davy seems to have received him rather coolly, and the impression at first conveyed by his reception appears to have been that Davy, having been recently married to a rich widow and at the same time created a baronet, was assuming rather offensive airs of superiority. However, after a call at the laboratory at the Royal Institution the visitor found Davy's attitude changed and he became the "*ganz vertraulicher kamerad*" expected. Davy gave him an introduction to the Astronomer Royal, Pond, who with the representatives of the Royal Society were assembled at Greenwich for the annual inspection of the Observatory. Pond introduced Berzelius to the company, among whom he made the acquaintance of many of the most important members. At the luncheon Berzelius found himself seated between the two secretaries, Wollaston and Thomas Young, while opposite were Sir Joseph Banks and Sir William Herschel; the rest of the guests were known to him by their more or less distinguished names. "This day," he says, "was one of the most memorable days in my life." On returning to London he found on his table a note from Davy thanking him for his remarks concerning the work on chlorine in which Davy had been recently engaged, and inviting his "brother of science" to his house for the next day. On arriving at Davy's house he was conducted by the French butler into a room in which breakfast was laid. But he was left to himself sufficiently long to admire the prodigious quantity of costly articles of luxury scattered round the room, which impressed the Swedish visitor very unfavourably. Lady Davy looked in while they were at breakfast, but did not take part in the meal. A long conversation ensued between the two chemists on scientific matters, especially concerning some of the statements in Davy's

Elements of Chemical Philosophy then issuing from the press. Berzelius expresses his satisfaction with the day he thus spent in Davy's company. He recognised in him a genius with unusually broad views, bold and independent in opinion, and shrinking before no difficulty in thinking out new methods, while ready to undertake any trouble in experiment. In regard to detail, Berzelius felt himself as much the superior as Davy was in all the rest. His amazing discoveries, based on profound considerations and carried out with inexhaustible perseverance, had raised him to the highest place of honour in the world of science. Berzelius thought that he had been tempted by his well-deserved title and the means now at his disposal to gratify his eagerness for distinction in another field in which his ambition would lead to disappointment. The man of science, he thought, would be honoured by the world until it was discovered that he unsuccessfully sought distinction in society, and then he would become a laughing-stock.

This was, in fact, the fate of the great British chemist, and it seems to have been the view of Berzelius that it was the recognition of these facts that led Davy to travel in other countries, where he might hope to meet with other triumphs. At the time of Berzelius' arrival in England Davy had already started on this new career and was under its intoxicating influence. Two days after the interview Davy left London and Berzelius saw him no more.

These remarks by the great Swedish chemist are severe but are not altogether without foundation. The simple life led in his own country, with small experience of such society as would be found in the fashionable centre of London, coupled with his own absorption in scientific pursuits, would probably have their influence on the view which such a man as Berzelius would take of Davy's circumstances and position in society. There is no doubt that Davy's marriage was in many respects unfortunate in tending to detach him from the pursuits which had been previously the chief occupation of his life.

What Davy thought of Berzelius can only be conjectured from the brief note composed by him among other notes on his contemporaries among foreign men of science, and already quoted (see p. 100).

Berzelius utilised this visit to England to form many acquaintances. Among the rest he mentions especially Alexander Marcet, who was at this time practising as a physician in London and at the same time lecturing on chemistry at Guy's Hospital. He was a Swiss who had settled in this country and was the husband of Mrs. Mary Marcet, well known as a writer of popular books on scientific subjects early in the century. After attending Marcet's lectures Berzelius acknowledged the important assistance he derived from seeing the experiments with which they were illustrated, and which he utilised in his own teaching after his return home. Berzelius also visited Windsor, but he records his feeling that the most important part of this occasion was the visit to Herschel, who showed him his great telescope at Slough. Guided by the secretary of the Royal Institution, Gullemard, he also went to Cambridge, "where," he says, "it was with a feeling of reverence I visited the room where Newton made the greater part of his splendid discoveries."

In company with Smithson Tennant he made an excursion to Bristol and the west of England, and accepted an invitation into Hertfordshire from Sir John Seebright. Thomas Young also took him to his country house near Worthing.

After an absence of five months he returned to Stockholm in November, 1812, full of gratification at the results of his journey, although the Royal Society was not in Session and Sir Joseph Banks' famous soirées were discontinued. He also carried back with him a large quantity of instruments and apparatus, and this enabled him to try many experiments which for lack of means he had been unable to accomplish previously.

On his arrival in Stockholm he was appointed by the King Director of the newly-established Academy of Agriculture with a salary, the appointment being conferred at first for three years, but subsequently for life.

Notwithstanding the exchange of friendly letters between Berzelius and Davy, they did not always agree on theoretical matters, and some of Davy's English friends went so far as to attribute to the former the appropriation of Davy's views. Berzelius wisely took no notice of these personal attacks.

Up to 1813 Berzelius in his researches on chemical proportions had not taken natural minerals into account. Silica had been previously classed with the earths, but Berzelius was now led to regard it as an acid and the silicates as salts analogous in constitution to alum. With this idea as guide Berzelius published a small treatise on chemical mineralogy which was soon afterwards translated into English and into German. Sixteen years later the mineral collections in the British Museum were arranged according to his system, and a little later the Royal Society awarded him the Copley Medal especially in consideration of his mineralogical researches. Many new Swedish minerals were examined, and with Gahn's aid Berzelius acquired great skill with the blowpipe, and published a text-book which found its way into almost every chemical laboratory.

In the year 1816 Berzelius was persuaded by Gahn, then over seventy years of age, to join him and some others in the purchase of a chemical works at Gripsholm. Berzelius was reluctant to yield, having already burnt his fingers as a consequence of attempting industrial enterprise; and as he himself remarks in his recollections, he had no talent for the industrial applications of science. The works were ultimately destroyed by fire, but one result of the examination of the sulphuric acid manufactured at Gripsholm was the discovery of the new element, selenium.

On the death of Klaproth in Berlin, Berzelius was invited to succeed him in the professorship of chemistry and other public offices connected therewith, and to settle the remuneration, but Berzelius declined this flattering offer on the very natural ground that he was unwilling to leave his own country, where he was free to devote himself to science.

During the winters of 1816-17 he had the satisfaction of teaching the young Crown Prince Oscar the elements of chemistry, and in the evening the royal family often visited his house, where he was able to show them physical and chemical experiments.

In later years his private laboratory was sought by both native and foreign students, and even mature chemists, among whom may be mentioned Sefström (teacher of chemistry at the Fahlun School of Mines), Arfwedson (member of the Swedish Academy) and C. G. Gmelin (afterwards Professor in Tübingen); and at a later time Heinrich and Gustav Rose, Mitscherlich, Magnus (all afterwards professors in Berlin), Mosander (afterwards professor in Stockholm) and Friedrich Wöhler, the friend of Liebig and afterwards professor at Göttingen.

About this time (1818-20) Berzelius' health, owing to hard work and the lack of fresh air and relaxation, began to suffer. He had for many years been subject periodically to pain in the head, which he rather curiously associated with the phases of the moon.

BERZELIUS' DWELLING HOUSE AND PRIVATE LABORATORY (1809-1819).

[Face p. 142]

But early in 1818 other symptoms appeared which seemed to suggest the propriety of a prolonged holiday. He accordingly accepted an invitation to visit Paris, and in July he set out, and passing through England, he renewed acquaintance with many of his old friends and visited many places in the course of the month spent in this country. He arrived in Paris on August 24th, and at once visited Berthollet, who received him with great friendliness and soon made him acquainted with Laplace, Cuvier, Gay-Lussac, Thénard, Chaptal, Dulong, Chevreul, Arago, Biot, Vauquelin, Ampère, Langier and other French men of science. He also met Alexander von Humboldt, then staying in Paris.

Berzelius attended many lectures and demonstrations by the most eminent chemists then in Paris and acknowledges that he received much valuable instruction, principally in the form of experiments suitable for lecture illustration. He also attended the meetings of the French Academy, of which he had been elected corresponding member in 1816. He became a Foreign Associate in 1822. On the other hand he found among the French chemists but little acquaintance with his work on chemical proportions and with his electro-chemical theory.

Berthollet invited him to spend some weeks in his house at Arcueil, the scene of so many famous assemblies, and while enjoying the society of the "noble patriarch of science," then approaching the end of his life, Berzelius was able to carry out in his laboratory a number of analyses of minerals. On his return to Paris he paid a visit to the gunpowder factory at Essonne where, under the guidance of Gay-Lussac, he gained a good deal of practical information. He was able also to forward to his own country a considerable quantity of instruments and apparatus which put him then in the position of being able toregard the equipment of his own laboratory as equal to that of any other in Europe. He left Paris in June, 1819. As he had seen but little of France, he arranged with his two former pupils, Arfredson and Almroth, a plan for the return journey which would take him through a part of France, through Savoy to Geneva and thence through Switzerland and Germany home.

In Geneva he met again his old friend Marcet, who had returned to his native place and had bought a property there. Here Berzelius met Theodore de Saussure, the elder De la Rive, Pictet, Prévost and other men of science. He visited Chamouni and many of the most interesting places in the neighbourhood, and, journeying northward, reached Tübingen, where he was received by his former student, C. G. Gmelin, now professor of chemistry in the university. A fortnight was spent in this district and a move was then made to Nürnberg, Erlangen and on to Dresden. In Berlin he stayed a fortnight, and again he visited many scientific friends and made many new acquaintances.

Berzelius, on his arrival in Stockholm, found that he had been elected secretary to the Academy, and he forthwith removed to apartments in the house of the Academy. As the result of his long expedition his health was much improved and his interest revived in all the new things in science which he had seen and learned in his travels. In a characteristic passage he expresses his thanks to Providence that he "was now one of the happiest of mankind." Unfortunately the pains in the head from which he had suffered previously returned, and he was advised to visit Carlsbad. The "cure" seems to have done more good than he expected. Of course he made many acquaintances and visited many places in the neighbourhood. Among the rest he called on Goethe at Eger, to whom he was introduced by Count Sternberg. Goethe received him with an air which seemed to indicate that he was not enchanted with the new acquaintance, and said nothing to him. However, he invited the visitors to lunch (Mittagessen), and afterwards they walked to the Kammerbühl. This is an extinct volcano but so small that one can monnt from foot to summit in a couple of minutes. A discussion between Goethe and Berzelius ensued on the origin and nature of the volcanic outburst, a subject on which Goethe had written a pamphlet some years previously. The poet was so much interested that he begged Berzelius to stay another day. The next day, after seeing his collection of minerals, Berzelius had to show him the use of the blowpipe, in which he became much interested and expressed the greatest regret that at his age (he was then over seventy) it would be impossible to become expert in the use of the instrument.

On the return journey homewards Berzelius paid a visit to his stepfather, Dr. Ekmarck, who received him with open arms. The meeting was an affecting one for both, as they were equally aware that it was the last. As a matter of fact, Ekmarck died the following day in his eighty-first year.

In the year 1824 Alexander Brongniart wrote to Berzelius announcing his intention of visiting Scandinavia for geological purposes, and Berzelius, being very desirous of returning some of

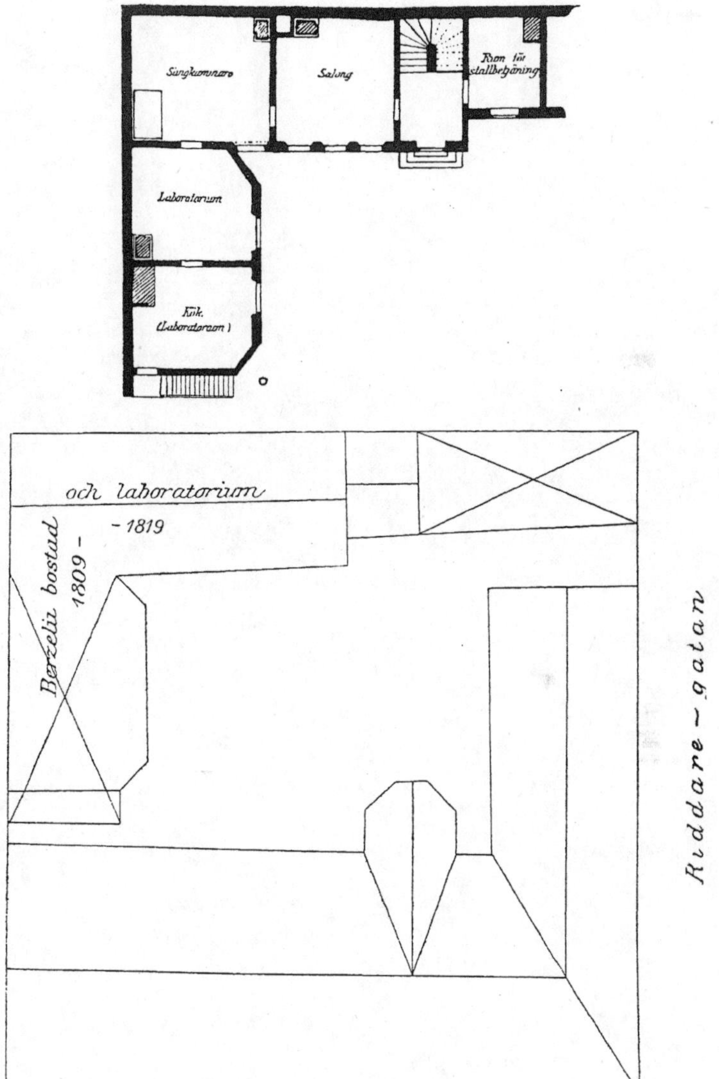

PLAN OF BERZELIUS' HOUSE AND LABORATORY. (First Floor).

[*Face p.* 144]

the attentions he had received when in Paris, determined to accompany him as interpreter. Accompanied by Wöhler, then a young man studying with him, he proceeded to Helsingborg, which is a port opposite the nearest point of Denmark. While in that town he received a letter from Sir Humphry Davy announcing that he was in Göteborg and expected to reach Helsingborg in a few days, at latest on July 29th. They waited till the evening of the 30th, and when about to start with Brongniart and Oersted, who had joined the party, they were informed that Davy had arrived. On hastening to the inn they learned from him that he had stayed on his way to Helsingborg two days in order to fish for salmon in the Laga river. "I will not conceal," says Berzelius, "that I took this plain excuse as evidence of disregard, though I did my best to conceal my feelings." (Davy was probably unaware of having given cause of offence. The following passage occurs in his diary of the journey: "At Helsingburgh I got a tolerable dinner, and a bed in an enormous room, probably the assembly room. Here I had an interview with Berzelius, whom I found in good plight, rather fatter than when I saw him twelve years ago.")

Toward the end of 1825 a Government Committee was appointed with a view to the reorganisation of national education, and it sat at intervals for the greater part of four years. Most people nowadays would agree with Berzelius in thinking it a small demand to require every University student on taking his degree, no matter in what faculty, to give evidence of a rudimentary knowledge of physical science and to be able to answer simple questions about our planetary system; to explain why the mercury mounts in a barometer, why liquids flow through a siphon, what goes on in the burning of a candle, why there is a draught in the chimney of a fireplace, or why ice swims on water, etc.; and though a century later than the time in which he was urging such reforms, it must be confessed that, while everyone should understand such phenomena occurring in daily life, not one in a hundred people know anything about them. Berzelius did not succeed in getting all he asked for, and he makes the remark that so long as those who stand at the head of the system of education in any country are ignorant of such important things no substantial reform is to be expected. This truth is as applicable in the present day and in every country as it was in the time of Berzelius and in Sweden.

In August, 1828, Berzelius, in company with Dr. Magnus, (afterwards Professor in Berlin), who had been working in his laboratory, started on a long round of visits through Germany, Belgium and Holland with the intention of taking part in the meeting in Berlin of the German "Naturforscherversammlung," an institution similar in aims to the British Association. At the request of the German Crown Prince, Berzelius was induced to give a popular lecture with experiments on a subject recently communicated to the Swedish Academy of Sciences. From the catalogue of his papers the subject was probably the bleaching action of the compounds of chlorine with bases. In 1830 he again attended the Naturforscherversammlung, this time in Hamburg.

In 1832 he retired from office as professor in the Karolinska Institute and was succeeded by Mosander, who had been both his pupil and assistant.

The outbreak of cholera in Stockholm in the summer of 1834 brought him other duties in the form of the Chairmanship of the committee appointed to deal with the situation. He escaped infection but he suffered again from the nervous disorder which had overtaken him in 1818. One of the results was that he lost temporarily not only his enthusiasm for scientific work but inclination for society. He began to feel himself lonely. His relations had advised him years before to marry and had painted in lively colours the miseries of a solitary old age with no one near to care for him. Previously he had refused to listen, but now he began to think their prophesies were about to be fulfilled.

He relates the fact that he had enquired many years earlier of a foreign scientific friend whose domestic happiness he had observed, whether, according to his experience, it was advisable for an active man of science to marry, and he received the following reply: "Although," said his friend, "I am as happy as only the father of a family can be, and the loss of my beloved wife would be a misfortune that I could scarcely bear, I believe that if I were now unmarried and had the same experience of life which I now possess I should certainly not marry except under the influence of an unconquerable passion." This, says Berzelius, determined him to give up the idea altogether, and with this in his mind he had grown up. But now he began to waver and thought it advisable in his state of indecision to con sult an intimate and experienced friend, Count Trolle-Wachtmeister. This was his opinion: "I suppose one can be quite

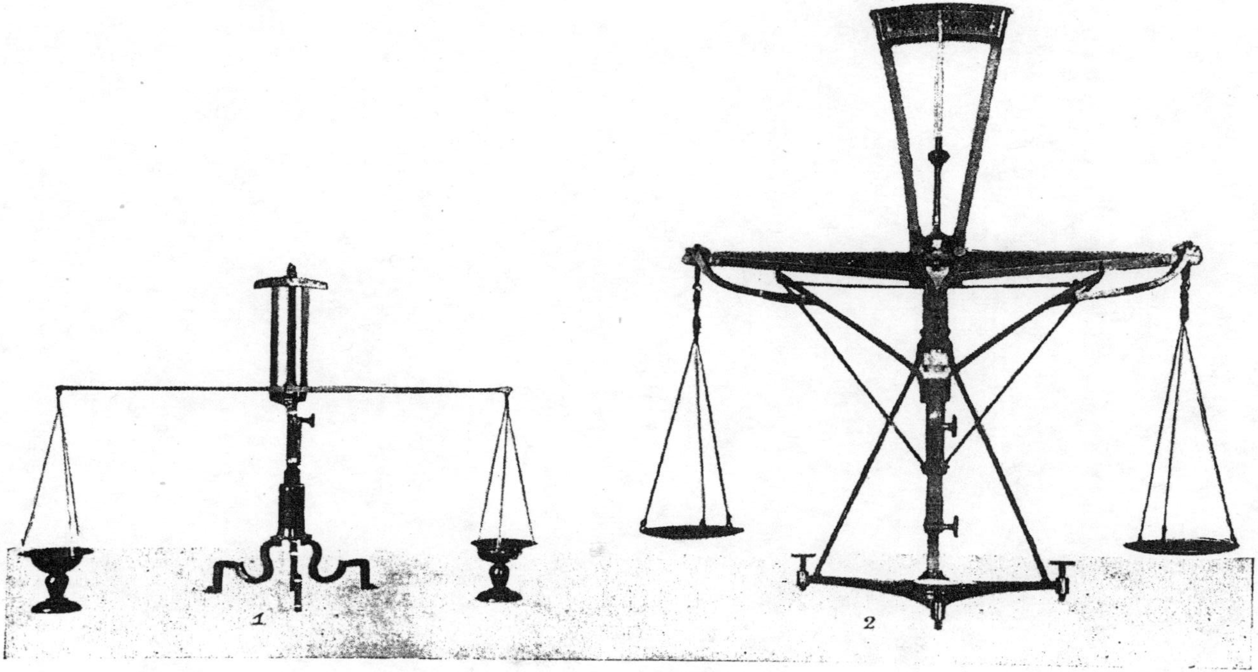

Balances used by Berzelius.
Figured in *The History of the Karolinska Institute.*

happy without entering into the married state, but he who has never experienced the happiness of having a beloved wife by his side knows nothing of the finest side of life. By a judicious choice it is not too late to enjoy this experience. To be perfectly happy a man should have a *chez soi* and he ought not to look for it outside his own dwelling." This enabled Berzelius to make up his mind, and a few days later he went to his old friend President Poppius and demanded the hand of his eldest daughter. Notwithstanding the great difference of age, the lady and her parents received him kindly. The whole proceeding is related with a *naïveté* which is extremely interesting and which seems to throw much light on the character of the great Swede.

Having decided on this important step, Berzelius determined to make a great effort to attain an improved state of health, and with this object in view he decided on paying another visit to Paris. While in the French capital he was introduced to King Louis Philippe, with whom he talked for an hour, the king using the Norwegian language and Berzelius the Swedish. The heir-apparent, Ferdinand, Duke of Orleans, told him that whatever he knew of chemistry he had learned from the French translation of his *Traité de Chimie.* This was evidently regarded by the author as a great compliment.

The French sculptor David used the opportunity to model a bust of Berzelius, and the work is now in the Karolinska Institute in Stockholm.

On his way homeward he stayed in Bonn during part of the meeting of Naturforscher, but being still unwell he went on with Wöhler to Cassel and thence to several other places, where he visited other scientific friends, ultimately reaching Stockholm in October.

On December 19, 1835, he was married, and the occasion was distinguished by his elevation to the peerage, while he received on the day of the wedding a letter from the King himself expressing in flattering words his recognition of the scientific labours to which his life had been devoted. His health remained variable, but in the summer he started with his wife to visit her relations in Denmark, where her uncle was Swedish ambassador. While in Copenhagen he was received by the Crown Prince and by the King of Denmark.

During the few years following this time Berzelius took an active part in the reorganisation of the Karolinska Institute in

Stockholm, a complete account of which has been published in the history issued by the Institute in the form of two illustrated jubilee volumes. He also took part in the formation of an association of Scandinavian naturalists (corresponding to the British Association), which met for the first time in 1839 at Göteborg.

The latter part of this interesting record was written in 1842. The author continued busily at work on the fifth edition of his treatise on chemistry, but the last two or three years of his life were attended by gradual decline of his bodily powers, and he died on August 1, 1848.

The laboratory where so much important work was done has always been a focus of interest. Not only was it the place where many discoveries were made by Berzelius himself, but it became the school where many of the young men who were destined to be the leaders in the next generation received the instruction in analysis and in methods of research which they needed but which were not to be found in the universities from which they came. Friedrich Wöhler, who was one of the most distinguished of these young men, gave to the Berlin Chemical Society in 1875, when he was an old man, a most lively and interesting story of reminiscences of his journey to Sweden in 1823, when in response to his appeal he was admitted by Berzelius to a place beside him. It is perhaps necessary to remind the reader that travel from one country to another was not accomplished so easily then as in later times. In the first place Wöhler arrived in Lübeck to find that he must wait six weeks for a ship, and he wisely occupied the period of waiting in the laboratory of an apothecary, F. Kindt, with whom he made acquaintance and who proved an enthusiast in the pursuit of natural knowledge. Between them they succeeded in producing a considerable quantity of potassium. When at last the little sailing-ship was loaded with cargo she started from Travemünde on October 25th, and after a rough but unusually short journey of four days she landed her four passengers at Dalarö, a small fort on the rocky coast of Sweden which it was then the custom to use as a landing-place to avoid the much longer sea journey to Stockholm. In consideration of the late period of the season Wöhler had brought provisions for three weeks, but now he quickly divided what was left among the sailors, and mounting an open vehicle drove on to Stockholm, which he reached the same night. After driving about some time he found quarters in a cellar, as in those days there were no inns in Stockholm.

In 1823 Berzelius occupied the apartments assigned to the secretary in the house of the Academy of Sciences. "As he led me," said Wöhler, "into his laboratory I was, as it were, in a dream, doubting whether it was really true that I was in this famous place. Adjoining the living-room, the laboratory consisted of two ordinary chambers with the simplest fittings; there was neither oven nor fume chamber, neither water nor gas supply. In one room stood two ordinary work-tables of deal; at one of these Berzelius had his working place, the other was assigned to me. On the walls were several cupboards with reagents which, however, were not provided very liberally, for when I wanted prussiate for my experiments I had to get it from Lübeck. In the middle of the room stood the mercury trough and glass-blower's table, the latter under one of the chimney-places provided with a curtain of oiled silk. The washing place consisted of a stone cistern having a tap with a pot under it. In the other room were the balances and other instruments, beside a small work-bench and lathe. In the kitchen, where the food was prepared by the severe old Anna, cook and factotum of the master who was still a bachelor, stood a small furnace and the ever-heated sandbath. Wöhler was at that time the only worker admitted into the private laboratory. There was no systematic instruction and Berzelius allowed each student to do what he liked, but he was there to give advice and help when needed. The first work undertaken by Wöhler on the advice of Berzelius was the quantitative examination of certain minerals, as he had had but little practice in the use of the balance. He was required to repeat his analyses until concordant results were obtained. When Wöhler had worked rather hastily, Berzelius uttered the stereotyped remark: "Doctor, das war schnell aber schlecht." (Doctor, that was quick but bad.) Among other subjects undertaken later was the investigation of cyanides and cyanates, which interested Berzelius very much, as it appeared important in connection with the question as to the nature of chlorine. It was at the time when its character as an elementary substance was becoming generally recognised, and Wöhler relates that Berzelius, who had been an active opponent of this theory, one day gave playful instruction to his female factotum. Anna in washing out a flask had remarked that it smelt of oxymuriatic acid, on which Berzelius said, "Anna, thou must no longer speak of oxymuriatic acid, henceforward thou must say chlorine."

If now an endeavour is made to review the chief features of Berzelius' life and work and to recall something of his personality, it must be admitted that Davy's brief estimate is justified, though incomplete. He undoubtedly devoted his whole time and energy to the pursuit of scientific research and to the development of theoretical chemistry. His autobiography shows how he concentrated all his thoughts on his chemical studies, and consequently that "his conversation was limited much to his own subjects." He had had a liberal education, and had it not been for the supreme attraction offered by the science of chemistry the fundamental principles of which were only beginning to be dimly perceived, he might have become a great surgeon or physician. In any career his immense capacity for work and his great activity of mind would have made him conspicuous. To appreciate the importance of his work in chemistry we have only to remember that thirty years before Faraday's researches on electrolysis Berzelius and Davy had both conceived electrical theories of chemical combination which differed from each other only in points of detail, but agreed in being both founded on definite results of experiment. Every student of chemistry ought to know that the symbols in universal use in the text-books for representing elementary atoms were contrived by Berzelius, and that the system arose directly from the introduction of the Atomic Theory by Dalton and the modification of the rather clumsy symbols used by him. The arduous investigations into the proportions in which elements combine to form compounds furnished the relative values which when corrected by reference to the Law of Volumes are spoken of as atomic weights. Berzelius made use of the Law of Specific Heats discovered in 1819 by Dulong and Petit, and of the facts of isomorphism discovered by E. Mitscherlich about the same time. By the study of the composition of the oxides and sulphides of metals and non-metals then known he placed on a firm basis the law of multiple proportions. It was Lavoisier who discovered the presence and the function of oxygen in oxysalts, such as sodium and copper sulphates, although he did not know that common salt and others like it did not contain oxygen in any form. This could not be settled before 1810, when Davy proved that chlorine is an elementary substance. But although Berzelius in his early days

J. BERZELIUS.
Late in Life.
(*From a Daguerreotype.*)

[*Face p.* 151]

shared the mistake about muriatic acid and its salts, so far as oxysalts were concerned he made one important observation when he discovered the relation between the amount of oxygen in the base of a normal salt and the amount of the acid (non-metallic oxide) to which this base was united. The system of notation he adopted, which was based on the binary idea of combination originated by Lavoisier, remained in general use till long after it was challenged by Gerhardt and ultimately exchanged for the unitary system. His work was done before the composition of many organic substances was known, and long before the idea of valency was recognised. Berzelius' views regarding the constitution of organic compounds had to be abandoned even before the end of his life.

Less than justice would be done to the memory of Berzelius. if it were not remembered that his influence as a teacher was extraordinary, and the fact that so many men who afterwards achieved high distinction resorted to his laboratory for that instruction which was not to be found in any of the universities. The autobiography tells of his early poverty and the struggle in which he was thus so long involved in the determination to complete his education and to devote his whole energies to the service of science. It also tells of his strong feelings of patriotism and the gratifying recognition of his successive steps in advance not only by the representatives of science throughout the world, but by the King and Government of his own country. Honours and rewards fell thick upon him, and in the portrait which is most familiar among chemists he is shown wearing some of the decorations the bestowal of which furnish proof of the high esteem in which science was held in the Sweden of his day.

Group VI

ELECTRO-CHEMISTRY

FARADAY (1791-1867)

CHAPTER XII

FARADAY

FEW persons who are familiar with the present position and operations of the Royal Institution are at the same time acquainted with its curious origin. The home of Davy and Faraday for a period of over fifty years, the reputation of the Institution is securely founded on the great discoveries in physical science which from its very beginning have been made within its walls. The complete history of the Institution was written by Dr. Bence Jones in 1871, and it is only necessary to recall the fact that it was founded by Count Rumford in 1799, at a time when the country, owing to the high price of corn, was passing through a period of great scarcity and there was much distress among the poor. A Society had been formed with the benevolent object of bettering the condition of the poor, and after conference with Rumford a committee of the Society reported that the Institution proposed by him would be extremely beneficial to the community. The proposals put forward were briefly indicated in the heading of a prospectus and were expressed in the following words:

"The proposals for forming by subscription, in the Metropolis of the British Empire, a Public Institution for diffusing the knowledge and facilitating the introduction of useful mechanical inventions and improvements, and for teaching by courses of philosophical lectures and experiments the application of science to the common purposes of life were these . . ." Then follow many details of the scheme for managing and carrying on the work of the Institution, which was designed to include an industrial school and a place of instruction for mechanics. In September, 1799, Dr. Garnett was engaged as lecturer and scientific secretary. The unfortunate state of his health and other circumstances led to his retirement in 1801, and early in that year Humphry Davy

MICHAEL FARADAY.

was appointed Assistant Lecturer and later, Lecturer on Chemistry. The course of events henceforward has been already described, and it is only necessary to add that the character of the work at the Institution was by the influence of Davy's genius transformed for all time. The Institution became the home, and almost the only centre in this country, of systematic scientific research. The brilliant discoveries of Davy made the Institution famous and if for a few years after his retirement the Institution did little for the advancement of science, one discovery was made which was destined in time to restore fully the reputation of the chemical laboratory. The appointment of Michael Faraday as laboratory assistant was an event of the first importance and accomplished under circumstances almost romantic.

Several congregations had been formed in different parts of England during the eighteenth century under the influence of Robert Sandeman, son-in-law of the Rev. John Glas, a Presbyterian clergyman in Scotland, and these congregations were known as Sandemanians or Glasites. The doctrine they professed assumed that the Bible alone contained all that was necessary to salvation, and that in all times and circumstances it should form the sufficient guide to Christian believers. Members were received into the Church on public confession of sin and profession of faith. Each church was governed by elders chosen among the members, from those who appeared to answer most nearly to the character of an elder as described in the New Testament. Among the brethren of such a congregation near Kirkby Stephen, in Westmoreland, the family of Faraday is first heard of, and here James Faraday was born in 1761. This James Faraday became a blacksmith, and in 1786 married Margaret Hastwell, a farmer's daughter of Mallestang, near Kirkby Stephen, and soon after their marriage the couple came to London. Four children were born of the marriage, and at Newington, in Surrey, the third child, Michael, was born on September 22nd, 1791. About 1796 the family removed to rooms over a coach-house in Jacobs Well Mews, Charles Street, Manchester Square, while the father continued his trade as a journeyman in Welbeck Street. He seems to have been in bad health for several years before he died, an event which occurred in 1810, soon after he had removed to 18, Weymouth Street, Portland Place. The home of little Michael was in Jacob's Well

Mews from the time he was five years old till he obtained employment in 1804, at the age of thirteen, as an errand boy, on trial for a year, in the service of Mr. George Riebau, a bookbinder and stationer, at 2, Blandford Street. A year later he was bound as an apprentice for seven years, and in his indentures the noteworthy statement occurs that "in consideration of his faithful service no premium is given."

There is but little to be told concerning the boy's early education beyond his own statement: "My education was of the most ordinary description, consisting of little more than the rudiments of reading, writing and arithmetic at a common day school. My hours out of school were passed at home and in the streets." But as soon as he entered Riebau's shop a new world of opportunity was opened to him and Faraday himself said, "Whilst an apprentice I loved to read the scientific books which were under my hands, and among them delighted in Marcet's *Conversations in Chemistry*, and the electrical treatises in the *Encyclopædia Britannica*. I made such simple experiments in chemistry as could be defrayed in their expense by a few pence per week, and also constructed an electrical machine first with a glass phial and afterwards with a real cylinder, as well as other electrical apparatus of a corresponding kind." He told a friend that Watts *On the Mind* first made him think, and that his attention was turned to science by the article "Electricity" in an encyclopædia he was employed to bind. His master also allowed him to go occasionally in the evening to hear lectures on natural philosophy given by Mr. Tatum at his house, 53, Dorset Street, Fleet Street. This Mr. Tatum seems to have been the founder of the City Philosophical Society, an association of some thirty or forty persons who met once a week in the evening for mutual instruction. Faraday became a member of this Society after the expiry of his apprenticeship, and it was through Mr. Tatum that he made the acquaintance of several young men of about his own age, with two of whom, Huxtable, a medical student, and Benjamin Abbott, a clerk in the city and a Quaker, he commenced a correspondence which was maintained during many years. The letters, of which most of Faraday's have been preserved, show the nature of the subjects in which they were mutually interested and in some degree reveal the character of the writers. They seem to show us the mind of an ingenuous and modest youth eager to escape from trade and "to enter," in his own words, "into the service of science which I imagined made its pursuers amiable and liberal."

In October, 1812, Faraday went as a journeyman bookbinder to a certain De la Roche, a French emigrant. His master's temper, however, rendered it impossible for him to remain long in this situation. He had in the spring of this year been taken by Mr. Dance, a customer of Riebau's and a member of the Royal Institution, to hear four of the last lectures of Sir Humphry Davy. He made notes of these lectures, with the addition of drawings, and then wrote to Sir Joseph Banks, President of the Royal Society. Naturally enough, nothing came of this attempt to escape from trade, but later in the year he applied directly to Sir Humphry Davy and sent him the amplified notes he had taken of his lectures. Davy's reply was as follows:

December 24th, 1812.

"To Mr. Faraday.

"Sir, I am far from displeased with the proof you have given me of your confidence, and which displays great zeal, power of memory and attention. I am obliged to go out of town and shall not be settled in town till the end of January: I will then see you at any time you wish. It would gratify me to be of any service to you; I wish it may be in my power.

"I am, Sir, Your obedient humble servant,

"H. Davy."

The promised interview took place, and Faraday's account of it contains the following passage: "At the same time that he thus gratified my desires for scientific employment he still advised me not to give up the prospects I had before me, telling me that science was a harsh mistress and in a pecuniary point of view but poorly rewarding those who devoted themselves to her service. He smiled at my notion of the superior moral feelings of philosophic men and said he would leave me to the experience of a few years to set me right on that matter." The end of the business was that on the recommendation of Davy he was appointed assistant in the laboratory of the Royal Institution at a salary of 25s. a week, with two rooms at the top of the house.

He had no sooner established himself in his new home than he made arrangements with his friend Abbott and others for carrying on in the evenings the education of which he felt in need, by means of the mutual improvement system, the meetings being held occasionally in his rooms at the Royal Institution.

Davy seems to have lost no time in making use of the services of his new assistant, for, in a letter to Abbott only six weeks after he entered on his duties, Faraday relates his experiences with "the detonating compound of chlorine and azote," and says, "I do it at my ease, for I have escaped (not quite unhurt) from four different and strong explosions of the substance." Both operators received damage from this dangerous compound on more than one occasion.

Letters to his friend Abbott at this time show that he was occupied daily in observations on the subjects of lectures and lecturers and the circumstances which contribute to success or failure. The Royal Institution provided material enough, for not only had Faraday to work in the laboratory but he was present at probably all the lectures given by the numerous outsiders who were invited to lecture on various subjects. A series of changes were, however, at hand, for Davy having married Mrs. Apreece in April, 1812, he ceased to lecture and was appointed Honorary Professor. About a year later he determined to take an extended tour on the Continent, and invited Faraday to accompany him and Lady Davy in the capacity of secretary and scientific assistant. On October 13, 1813, the whole party started from London, arriving at Plymouth on the 15th. Davy had engaged a valet in London, but the man withdrew at the last moment and no other servant could be found either before starting or later in Paris or other towns they visited. The result was that on many occasions Faraday was called upon to make arrangements which would otherwise have constituted part of the duty of the valet and which led to much annoyance for Faraday. His position is sufficiently indicated by a few lines in a letter to his friend Abbott after the travellers had been together upwards of a year. He says: "I should have but little to complain of were I travelling with Sir Humphry alone, or were Lady Davy like him; but her temper makes it oftentimes go wrong with me, with herself and with Sir H."

Faraday kept a journal during the whole of the journey, and many passages are interesting as illustrating the inconveniences of travel in Europe more than a hundred years ago, beside showing his habit of careful observation. He worked at the French and Italian languages and was evidently much influenced by the desire for improvement expressed in letters both before and after departure from his native country. In one of his first letters from Paris to his mother he says: "The first and last thing in my mind is England, home and friends. . . . Whenever present circumstances are disagreeable I amuse myself by thinking of those at home. In short, when sick, when cold, when tired, the thoughts of those at home are a warm and refreshing balm to my heart . . . these are the first and greatest sweetness in the life of man." And in another place: "I have learned just enough to perceive my ignorance, and ashamed of my defects in everything, I wish to seize the opportunity of remedying them. . . . Added to which the glorious opportunity I enjoy of improving in the knowledge of chemistry and the sciences continually determines me to finish this voyage with Sir Humphry Davy."

Throughout his absence from England Faraday kept a journal, which contains some amusing passages; for instance, his account of the examination of the travellers and their baggage on arrival in harbour at Morlaix, the landing of the travelling carriage, the equipment and appearance of the postillion and the interior of the hotel at Morlaix. What excited his interest almost more than all the rest was that during a breakdown on the road he saw for the first time in his life a *glowworm.* It took about a week to drive to Paris, and Faraday's sensations on finding himself in the enemy metropolis seem to have been mixed and far from affording full satisfaction. The party remained in Paris just two months, and proceeding by Montpellier and Nice they crossed the Col di Tenda and reached Turin on Tuesday, January 22, 1814. On their arrival in Turin the carnival was still in progress and the behaviour and occupation of the crowd of citizens was described by Faraday in his journal with a mixture of amusement and contempt. He records a visit to the opera at Genoa, which they reached a few days later, and ends the entry in his diary with the remark, "the performance to me very tedious."

While at Genoa he had an opportunity of observing the phenomena exhibited by three waterspouts, appearing together from the same stratum of cloud. The next place visited was Florence, and here Faraday assisted Davy in the experiment of burning a diamond in oxygen by the aid of the Duke's great lens for concentrating the sun's rays. The whole experiment is described with interesting details: "A glass globe containing about 22 cubical inches was exhausted of air, and filled with very pure oxygen procured from oxymuriate of potash; the diamond was supported in the centre of this globe by a rod of platinum, to the top of which a cradle or cup was fixed, pierced full of holes to allow free circulation of the gas about the diamond. The Duke's burning-glass was the instrument used to apply heat to the diamond. It consists of two double convex lenses, distant from each other about 3½ feet; the large lens is about 14 or 15 inches in diameter, the smaller about 3 inches in diameter. The instrument is fixed in the centre of a round table, and is so arranged to admit of elevation or depression, or any adjustment required, at pleasure. By means of the second lens the focus is very much reduced, and the heat, when the sun shines brightly, rendered very intense. The instrument was placed in an upper room of the museum; and having arranged it at the window, the diamond was placed in the focus and anxiously watched. The heat was thus continued at intervals for three-quarters of an hour (it being necessary to cool the globe at times), and during that time it was thought that the diamond was slowly diminishing and becoming opaque. Now we had only a partial spectrum, for the upper part of the window obstructed the sun's rays; but having sunk the whole of the apparatus it was again exposed and a very strong heat obtained. On a sudden Sir H. Davy observed the diamond to burn visibly, and when removed from the focus it was found to be in a state of active and rapid combustion. The diamond glowed brilliantly with a scarlet light inclining to purple, and when placed in the dark continued to burn for about four minutes. After cooling the glass, heat was again applied to the diamond and it burnt again, though not nearly so long as before. This was repeated twice more, and soon after the diamond became all consumed. This phenomenon of actual and vivid combustion, which has never been observed before, was attributed by Sir H. Davy to the free access of air. It became more dull as carbonic acid gas formed and did not last so long." Similar experiments were made on several successive days, and the gas formed was examined and proved to consist of nothing except carbonic acid gas. The experiment was then repeated with the substitution of chlorine for oxygen, but no change could be observed in the diamond and no combination of carbon and chlorine could be effected. "Having finished these experiments," Faraday says, "we bade adieu for a time to the Academy del Cimento and prepared to depart for Rome."

After a few days in Rome the party proceeded to Naples and on Friday, May 13th, the entry in the diary records that "Mount Vesuvius was the employment of to-day and fully rewarded the trouble and fatigue attendant upon seeing it." The next day was also devoted to the ascent, the object being to see the mountain by night, and the journal winds up the day by the note: "Got home by half-past eleven o'clock, highly pleased and satisfied with the excursion." Returning northward, the travellers reached Milan, and on Friday, June 17th, "saw M. Volta, who came to see Sir H. Davy, a hale elderly man, bearing the red ribbon, and very free in conversation." In the course of a letter written from Geneva in August to his friend Abbott, Faraday enters in characteristic fashion into a long dissertation on the qualities of the people he had been brought into contact with in his travels, and his ideas on the subject of the French and Italian languages. "During the time I have passed from home many sources of information have been opened to me and many new views have arisen of men, manners and things both moral and philosophical. The constant presence of Sir Humphry Davy is a mine inexhaustible of knowledge and improvement; and the various and free conversation of the inhabitants of those countries through which I have passed has continually afforded entertainment and instruction." A month later, in writing to the same friend, he is in a less satisfied mood, for while acknowledging the advantages of travel, he says: "In the first place, then, my dear B., I fancy that when I set my foot in England I shall never take it out again; for I find the prospect so different from what it at first appeared to be, that I am certain if I could have foreseen the things that have passed I should never have left London." He goes on to say: "The little knowledge I have gained in languages makes me wish to know more of them, and the little I have seen of men and manners is just enough to make me desirous of seeing more; added to which the glorious opportunity I enjoy of improving in the knowledge of chemistry and the sciences continually determines me to finish this voyage with Sir Humphry Davy. But if I wish to enjoy those advantages, I have to sacrifice much; and though those sacrifices are such as an humble man would not feel, yet I cannot quietly make them. Travelling, too, I find is almost inconsistent with religion (I mean modern travelling), and I am yet so old-fashioned as to remember strongly (I hope perfectly) my youthful education." It seems, however, that even Faraday could not resist the influence of the gaiety by which he was sometimes surrounded. The winter of 1814-15 was passed in Rome, and we learn from the diary and letters that during the carnival he went on several occasions to the masked balls which are a feature of the carnival season. February 6th has the entry: "Went to this morning's masked ball in a domino and found it very full; as no one knew me, at least for some hours, I amused myself a good deal with such as I was acquainted with. I stopped there till daylight, and then came home." On the 11th we find a different sort of entry, more consistent with the character of the philosopher: "Experiments at home on the new compound of oxygen and chlorine which Sir Humphry discovered a few days ago." A letter to his mother, dated April 16, 1815, announcing his expected return home contains the brief explanation: "I am not acquainted with the reason of our sudden return; it is, however, sufficient for me that it has taken place. We left Naples very hastily, perhaps because of the motion of the Neapolitan troops, and perhaps for private reasons. We came rapidly to Rome, we as rapidly left it. We ran up Italy, we crossed the Tyrol, we stepped over Germany, we entered Holland, and we are now at Brussels, and talk of leaving it to-morrow for Ostend; at Ostend we embark, and at Deal we land on a spot of earth which I will never leave again."

A fortnight after his return to England Faraday was engaged at the Royal Institution as assistant in the laboratory and superintendent of the apparatus at a salary of 30s. a week, with apartments. His duties naturally brought him into contact with not only Sir H. Davy but Professor Brande and many other persons interested in science or occupied in its pursuit: they also afforded daily opportunities for handling and using apparatus and for making experiments, and so supplied him with the means of making progress in manipulation and in the furnishing of his own mind. He soon resumed his association with the City Philosophical Society, and in January, 1816, he began a course of lectures on chemistry to the members. He seems to have kept a commonplace book in which he recorded not only observations of his own and memoranda of all kinds, but notes of lectures given at the Royal Institution by Brande, who had succeeded Davy as Professor.

His first paper appeared in 1816 in the *Quarterly Journal of Science* on the "Analysis of the native caustic lime of Tuscany." In the volume of his *Experimental Researches in Chemistry and Physics* published long years afterwards he explains that "Sir Humphry Davy gave me the analysis to make as a first attempt in chemistry, at a time when my fear was greater than my confidence, and both far greater than my knowledge; at a time also when I had no thought of ever writing an original paper on Science." The same journal contains in the following year no fewer than six other papers from his pen. One of these on the escape of gases through capillary tubes was the forerunner of work done by Graham thirty years later.

The smooth course of his life at the Royal Institution was but little interrupted. In the summer of 1817 he spent a month with his friend Huxtable near South Molton, on the border of Exmoor. "At the Narracote sheep-shearing Mr. Faraday took part in the conviviality of the evening with much apparent interest and good humour" is the note made by Huxtable. At this time the experiments on flame were made by Sir H. Davy, which led to the invention of the *Safety Lamp*, and Faraday communicated papers to the *Quarterly Journal* on the subject. He lectured again and again at the Philosophical Society, and his note-books were filled with remarks on an immense diversity of subjects, among which in 1818 occur frequent references to electrical experiments. The papers published at this time show that he was pursuing many subjects but no one which as yet claimed continuous attention.

In the summer of 1819 he went for a three weeks' holiday on a walking tour in Wales. His journal gives evidence of his enjoyment of scenery and observation of natural phenomena. His remarks often display a quiet humour and also remind us how different were the conditions of such a ramble among the Welsh mountains a hundred years ago as compared with those of to-day. He begins: "The Regulator is an excellent coach. I mounted the top of it at the White Horse, Piccadilly, about a quarter past five on Saturday morning, July 10, and it set me down in Bristol about ten o'clock the same evening." From Bristol Faraday and his companion (it is uncertain who this was, perhaps Huxtable) crossed to Cardiff and proceeded thence to Merthyr and Neath, afterwards into Brecknockshire, and their further rambles included the ascent of Cader Idris, on the summit of which they were overtaken by a thunderstorm. Ultimately they got to Bangor and wound up at Llangollen.

Sir Humphry Davy was at this time in Italy and corresponded with Faraday on the subject of the papyri from Naples, but in the course of the next few months a step was taken which, while in no degree interfering with his work as a man of science, made Faraday a happy man for the rest of his life. This was his engagement to Sarah, third daughter of Mr. Barnard, of Paternoster Row, an elder of the Sandemanian Church in which Faraday himself had been brought up. In May, 1821, Faraday was appointed superintendent of the house and laboratory at the Royal Institution, and being thus in a position to provide a home for his wife, they were married on June 12, 1821. It was characteristic of the man that the wedding-day was considered "just like any other day," and in a letter to his sister-in-law, Mrs. Reid, he says: "There will be no bustle, no noise, no hurry occasioned even in one day's proceeding. In externals that day will pass like all others, for it is in the heart that we expect and look for pleasure."

His scientific work included papers on a variety of subjects, among them a series of experiments undertaken in conjunction with J. Stodart, a scientific instrument-maker, with a view to the improvement of steel, which, however, led to no great results. But the two or three years following were marked by several events which brought Faraday into a new position before the world. In the first place he began experiments in the field already laid open by the discoveries of Oersted, of Copenhagen, and by the French physicist Ampère, regarding the mutual action of magnetism and electric currents. He also began experiments on the action of pressure and cold on gases, whereby first chlorine, and subsequently many other gases were reduced to liquids, and his results were published in the *Philosophical Transactions of the Royal Society*. In 1824 he was elected a Fellow of the Society. Faraday's name was proposed by Richard Phillips, Wollaston, Children, W. Babington and Sir W. Herschel, but Davy as President and Brande as Secretary would be precluded by custom from adding their names to the certificate. This is not sufficient to explain the fact that, influenced by motives which it is very difficult to account for, Davy, who had previously been uniformly kind and friendly to Faraday, opposed his election. However, it is equally certain that the feelings which thus

FARADAY'S LABORATORY AT THE ROYAL INSTITUTION.

From Bence Jones' *Life of Faraday*.

interrupted the complete harmony previously existing for so many years soon passed away. In February, 1825, the managers, acting on the proposal of Sir H. Davy, promoted Faraday from the position of assistant to that of Director of the Laboratory, under the superintendence of the Professor of Chemistry. His experiments in connection with electro-magnetic phenomena were not allowed to proceed without interruption, for in 1824 the Council of the Royal Society appointed a committee for the improvement of glass for optical purposes, and Faraday was put on this committee. Shortly afterwards a sub-committee, consisting of Herschel, Dollond and Faraday was appointed to carry out experiments. The research was a long and laborious one. It did not end in the improvement of telescopes, but the dense glasses produced became important in connection with Faraday's diamagnetic and magneto-optical researches somewhat later. Much of the work required the erection of a furnace at the Royal Institution, and an assistant was engaged, Sergeant Anderson, of the Royal Artillery, who became a notable figure familiar to all frequenters of the lectures for more than five and thirty years.

The long preparatory course of experimental work led ultimately to the electrical researches which have rendered the name of Faraday famous for all time. Beginning in 1824 with the knowledge that as a current of electricity affects a magnet, there must be a reaction of the magnet on the current, and the one would, if free, revolve round the other, in 1831 he discovered electro-magnetic induction, and so rendered possible the invention of the induction coil and magneto-electric machines, the forerunners of the modern dynamo, by which electric currents are generated and rendered applicable on a large scale to the practical purposes with which all are now familiar. And this is how he wrote about his doings to his friend Richard Phillips; the letter is dated Brighton, November 29, 1831: "We are here to refresh. I have been working and writing a paper that always knocks me up in health, but now I feel well again and able to pursue my subject, and now I will tell you what it is about. The title will be, I think, 'Experimental Researches in Electricity': I. On the Induction of Electric Currents; II. On the Evolution of Electricity from Magnetism; III. On a new Electrical Condition of Matter; IV. On Arago's Magnetic Phenomena." And then the writer gives to his friend a short account of his principal experiments. But the remark quoted at the beginning serves as a warning of what was to follow. Faraday was now forty years of age, and for nearly twenty years had been constantly occupied in researches either under the direction of his famous master, Davy, or his own, but all requiring constant thought. The result was over-fatigue of the brain accompanied by symptoms which are not recorded as to the early stages, but in a few years led to fits of loss of memory and giddiness. But he worked on, with occasional holidays such as the one at Brighton just referred to, and a tour in France and Switzerland which he made in company with his wife and her brother in 1835.

About this time Faraday became occupied with the question of electric conductivity through solids and liquids, which led in the end to the great laws of electro-chemical decomposition which lie at the foundation of so much in chemical theory. By 1834 he proposed the new terminology which has been familiar for so long in chemical language. He would have avoided the word *current* if he could. In place of the word *pole* he substituted the neutral term, *electrode*. He applied the term *electrolyte* to every substance which can be decomposed by the current, and the act of decomposition he called *electrolysis*. The positive electrode he called the *anode* and the negative the *cathode*, while the constituents of the decomposing electrolyte he called the *ions*. The youngest student of chemistry in these latter days is familiar with all these terms, and he cannot go far without discovering that the borderland between physics and chemistry is occupied with discussions relating to the existence of the ions, their actions, reactions and behaviour under various conditions. And it is no wonder that Faraday began to feel the strain of incessant work, for with all the research going on he was still lecturing at the Royal Institution. Since 1829 he had been also Lecturer at the Royal Military Academy at Woolwich, and had accepted the appointment of Scientific Adviser to the Trinity House in 1836. The work chiefly on lighthouse illumination engaged his attention for thirty years. Of course honours of all kinds were showered on him, and in 1835 a pension was granted to him by the British Government, in which Lord Melbourne was Prime Minister. In 1839 and 1840 less original work was done owing to the more frequent feelings of fatigue.

The most important event in Faraday's life in 1840 was his

election as an elder by the Sandemanian Church. This involved preaching on alternate Sundays during three and a half years. It represented no new experience, as he had been frequently called upon to exhort the brethren, but however consonant, as it was, with Faraday's own feelings and principles, it left still less time for the rest and recreation necessary for the relief of his overworked brain.

In 1841 it became necessary to call a halt, to stop all lecturing and experimental work, and for four years his life had to be regulated accordingly. In the summer of this year the Faradays, with Mr. and Mrs. George Barnard, went for three months to the Rhine and Switzerland. They rested for some time at Thun, and here Faraday made excursions on foot which gave evidence of considerable activity and physical energy. On one of these over the Gemmi Pass back to Thun he records in his diary forty-five miles in ten and a half hours, excluding two hours of rest: "So that I think my strength cannot be bad or my reasoning (?) very insufficient. I would gladly give half this strength for as much memory, but—what have I to do with that? Be thankful."

In 1845 Faraday resumed work on the condensation of gases, and with improved appliances reduced many more to the liquid state, leaving hydrogen, oxygen, nitrogen, carbonic oxide, marsh gas and nitric oxide, which only yielded to experiments in other hands long afterwards, in the light of Andrews' discovery of critical temperatures.

In the autumn of 1845 he made the remarkable discovery of the influence of a magnetic field of force on polarised light, and he pursued a series of experiments on the division of all kinds of matter into those which are attracted (magnetic) and those which are repelled by the magnetic poles (diamagnetic bodies). His further electrical researches were published in the *Philosophical Transactions*. For these discoveries the Royal Society in 1846 awarded to him both the Rumford and the Royal Medals, an unusual double honour justified by the unusual occasion.

The experiments which led to the discovery of what he called magne-crystallic action followed naturally from his preceding work, but though Faraday continued to pursue his researches at the Royal Institution and to lecture and to visit and report on lighthouses, remarks in letters to his wife and to his friends show that the strain which had been imposed so long on his physical powers, especially in respect to memory, had not been permanently relieved by the protracted holiday in 1841 and the three years following.

About the year 1853 the phenomena exhibited in "table-turning" attracted a great deal of attention, and Faraday was induced to make an investigation into the subject. The result was a letter to the *Times*, and at the Christmas lectures on Voltaic Electricity he concluded with remarks which sufficiently show his attitude toward the then excited condition of the public mind. These words of advice are just as valuable now as they were in Faraday's time and as they must ever remain hereafter. He said: "In conclusion I must address a few words to the intending philosophers who form the juvenile part of my audience. Study science with earnestness—search into nature—elicit the truth—reason on it and reject all which will not stand the closest investigation. Keep your imagination within bounds, taking heed lest it run away with your judgment. Above all let me warn you young ones of the danger of being led away by the superstitions which at this day of boasted progress are a disgrace to the age, and which afford astonishing proofs of the vast floods of ignorance overflowing and desolating the highest places."

Faraday was now over sixty years of age, and 1855 saw the end of that wonderful series of researches and discoveries in electricity and magnetism which were begun in 1831. It is worth while to recall the circumstances under which this amazing amount of work was done. The Royal Institution had been kept alive by the lectures given by Faraday. He had no grant from the Royal Society or from Government, except the pension in 1835; the Institution could afford to give him only £100 a year, to which the Fullerian professorship added about £100 more.

In 1858, through the thoughtful kindness of the Prince Consort, the Queen offered Faraday the use of a house on Hampton Court Green, and here he and his wife lived to the end of his life. From this time onward increasing infirmity, chiefly in the form of loss of memory, led him to escape as much as possible from engagements of all kinds. He was able to examine and report on some of the lighthouses for the Trinity House and had the satisfaction of seeing the magneto-electric light at the South Foreland and Dungeness. In 1862 he was examined by the Public School Commissioners, and among his most memorable answers was the expression of his surprise at the opposition which had been raised to the introduction of natural knowledge into schools. "That the natural knowledge which has been given to the world in such abundance during the last fifty years," he said, "should remain untouched, and that no sufficient attempt should be made to convey it to the young mind growing up and obtaining its first views of these things, is to me a matter so strange that I find it difficult to understand: though I think I see the opposition breaking away it is yet a very hard one to be overcome. That it ought to be overcome I have not the least doubt in the world."

In 1865 he made his last report to the Trinity House, on the St. Bees Light. He also relinquished all duties connected with the Royal Institution. The remainder of his life was spent peacefully at Hampton Court, and, ever conscious of the approaching end, his reply to a friend from London, who came to enquire after his health, was, "Just waiting." He passed away quietly in his chair on August 25, 1867, and was buried in Highgate Cemetery.

The story of Faraday's life and work would be incomplete without reference to the extensive correspondence carried on throughout his life with his relations and early friends, such as his friends Abbott and Huxtable already referred to, and with scientific men, especially abroad. The letters, of which many have been preserved, give a better view of his character than could be derived from his daily work and other sources of information. When in Geneva with Sir Humphry and Lady Davy, in 1818, Faraday made the acquaintance of Professor G. de la Rive, with whom a correspondence was begun soon after his return to England and in later years was continued with his son, Professor Auguste de la Rive, altogether nearly fifty years. These letters contain not only mention of the work in which he was engaged and of his successive discoveries, but they demonstrate the cordial intimacy which existed between the writers.

Another correspondent with whom letters were exchanged during many years was Professor Schönbein, of Bâle, the discoverer of ozone. There are many other letters, among those which have been preserved, from Faraday to various persons and to Faraday from many of the most famous scientific men in other countries, all testifying to the extraordinary interest excited throughout the world in his work in every direction, and admiration of the successive discoveries proceeding from his laboratory.

One letter addressed by Faraday to a lady who had expressed a wish to become his pupil and assistant contains a passage which shows more clearly than anything which could have been written about him, the position which he had taken up early in life with respect to religion. It is interesting in view of the fact that the views he had adopted penetrated into his daily life and influenced all his actions. The passage is as follows: "You speak of religion, and here you will be sadly disappointed in me. You will perhaps remember that I guessed, and not very far aside, your tendency in this respect. Your confidence in me claims in return mine to you, which indeed I have no hesitation to give on fitting occasions, but these I think are very few, for in my mind religious conversation is generally in vain. There is no philosophy in my religion. I am of a very small and despised sect of Christians, known, if known at all, as *Sandemanians*, and our hope is founded on the faith that is in Christ. But though the natural works of God can never by any possibility come in contradiction with the higher things that belong to our future existence, and must with everything concerning Him ever glorify Him, still I do not think it at all necessary to tie the study of the natural sciences and religion together, and in my intercourse with my fellow-creatures that which is religious and that which is philosophical have ever been two distinct things."

On reviewing the career of Faraday on the one hand, and of Davy, his great predecessor, on the other, it is impossible to avoid recognising the contrast between the characters of the two men. Agreeing in their ardent pursuit of new knowledge in experimental skill, in acute observation and scientific deductive and inductive reasoning, Davy was distinguished by his poetic nature, the eagerness with which he courted fame, and the evident pleasure which he derived from intellectual society. Faraday from first to last preferred the more humble position of a simple enquirer after natural knowledge, with a remarkable indifference alike to distinctions and rewards and to the position commonly accorded to wealth. Davy's discovery of potassium and sodium supplied a new starting-point for the whole science of chemistry, while his invention of the *Safety Lamp* has been the means of saving thousands of human lives. Faraday's work on electricity lies at the foundation of all modern knowledge of that subject and has given to the world a means of controlling and directing physical energy previously undreamed of. Their joint services to humanity are of inexpressible value and their names will be ever, both jointly and separately, held in remembrance, not only in their own country but throughout the civilised world.

Group VII

AVOGADRO (1776-1856)

CANNIZZARO (1826-1910)

CHAPTER XIII

AVOGADRO

"ESSAI d'une manière de déterminer les masses relatives des molécules élémentaires des corps, et les proportions selons lesquelles elles entrent dans ces combinaisons." This is the title of the famous essay published in the *Journal de Physique* for 1811 (vol. lxxiii., pp. 58-76) in which what is known as the *Law of Avogadro* was enunciated. The centenary of this publication has been celebrated by the issue of a handsome volume under the title *Opere Scelte di Amedeo Avogadro* by the Royal Academy of Sciences of Turin. It is prefaced by a historical introduction from the pen of the late Professor Icilio Guareschi, and from it may be gathered all that is now recoverable concerning the life and personality of the author of the essay. The following pages contain a somewhat abbreviated account.

AMEDEO AVOGADRO, or, to give him his full title, Lorenzo Romano Amedeo Carlo Avogadro di Quaregna e di Cerreto, was born in Turin, August 9, 1776. The family name Avogadro appears to have originated in *De Advocatis*, which by gradual changes in the language became successively *Advocarii*, *Avocarii*, *Avogadri*, in reference to the legal functions discharged in ancient times in connection especially with ecclesiastical business, and which became hereditary. There were two divisions of the family connected respectively with Trevise and Vercelli. The father of the physicist, Cavaliere Philippo Avogadro, a member of the latter division, died in 1812; his mother was Anna Vercellone di Biella, a small place in Lombardy, near the property of the Avogadri.

Amedeo Avogadro obtained in 1789 the licentiate in philosophy, and in 1792 the baccalaureate in jurisprudence, ultimately attaining the doctorate in ecclesiastical law. For some years he was engaged in the practice of the legal profession,

From the Bronze Bust by P. Canonica at Turin.

[*Face p.* 170]

but from about 1800 he began the serious study of mathematics and physics, and in 1809 was appointed professor of Physics at the Royal College or Academy at Vercelli. His first original work, done in association with his brother Felice, on an electrical subject, was presented to the Academy of Sciences in Turin, September 20, 1803.

In 1820 the first Italian chair of mathematical physics was instituted by Victor Emanuel I. in the University of Turin, and Avogadro held this professorship till the end of 1822, when it was suppressed for political reasons. Avogadro received the title of Emeritus Professor with a small annual salary of 600 lire, and continued to occupy himself with his favourite scientific subjects in connection with physics and chemistry. His most important memoirs were published between 1811 and 1821, but others still interesting were produced during the following thirty years.

In 1832 the chair of mathematical physics was restored, but was occupied by the French physicist Cauchy during the first two years, when it was again given to Avogadro, who held it down to 1850. He then retired, and was succeeded by his pupil, Felice Chiò.

Avogadro married Donna Felicita Mazzi, of Biella, by whom he had six sons, of whom two rose to positions of distinction: Count Luigi as General in the Italian army, and advocate Felice, who became President of the Court of Appeal. Avogadro himself led a busy life and filled many public offices connected with national statistics, with meteorology, with weights and measures, and in 1848 he became a member of the Superior Council on Public Instruction. He cultivated not only Italian but the Greek and Latin literatures, and was familiar with English and German languages. He resembled Scheele and Dalton in the industry and modesty of his life. He cared nothing for prominent position or honours. At the Scientific Congress at Turin in 1840 he was not even nominated a Vice-President. He led, indeed, the life of a philosopher of the ancient type, occupied wholly with his studies, while not forgetting his duties as a citizen and father of a family. One of the consequences of this retiring habit was that the fundamental law associated with his name was not immediately appreciated and adopted. It must be remembered, however, that at the time (1811) it was enunciated, chemists had not become familiar with the atomic doctrine so recently promulgated by Dalton and his supporters. The word *molecule* was then unknown in scientific literature, and being freely used by Avogadro in his first memoir (*Journal de Physique*, July, 1811, Paris) it probably led to some confusion in the minds of his readers, and induced them to attribute to him a confusion which did not exist.

It would be beyond the purpose of this book to reproduce the whole of Avogadro's essay of 1811, which has already appeared in English among the *Alembic Club Reprints* (No. 4), and of which the essential statement is to be found in every modern chemical text-book. It is only necessary to quote here the note which has been appropriately added by Professor Walker to the English version. It is as follows: "Avogadro has been accused of inconsistency in his use of the term 'molecule,' but a careful perusal of his paper will show that he uses it with its qualifying adjectives quite consistently as follows:

"*Molécule* (translated 'molecule') without qualification means in modern chemical phraseology either *atom* or *molecule*.

"*Molécule integrante* (translated 'integral molecule') means *molecule* in general, but is usually applied only to compounds.

"*Molécule constituante* (translated 'constituent molecule') is employed to denote the *molecule* of an elementary substance.

"*Molécule élémentaire* (translated 'elementary molecule' stands for the *atom* of an elementary substance."

The reader may be reminded that it was Avogadro who showed for the first time that the molecules of many elements are composed of more than one atom and consequently that certain cases of combination between elements must be represented, as is now universally customary, by symbols denoting a double decomposition. It is only necessary to recall the two cases of combination of hydrogen with chlorine, and hydrogen with oxygen, which are formulated in the following manner:

$$\mathrm{HH} + \mathrm{ClCl} = \mathrm{HCl} + \mathrm{HCl} \text{ and}$$
$$\mathrm{HH} + \mathrm{HH} + \mathrm{OO} = \mathrm{HOH} + \mathrm{HOH}.$$

In these and all similar cases the formula expresses the same volume of element or resulting compound so long as it remains in the state of gas and at the same temperature and pressure.

It is deplorable to reflect that justice was so long denied to Avogadro and his great discovery. In 1814 the French physicist Ampère addressed a letter to Berthollet in which he "adopted," to use Avogadro's own words, the ideas which had been put forward three years earlier. For a long time the hypothesis or law was by the French associated with the name of Ampère, and it was only long after Avogadro's death in 1856 that the chemical world was forced in deference to the representations of Cannizzaro, to adopt the principle and recognise the merit of his great compatriot, and the reader is therefore referred for amplification of this brief account of Avogadro's views to the pages in which Cannizzaro's exposition is described and concisely discussed.

At the death of Berzelius in 1848 theoretical chemistry was in a position of great confusion, from which it took nearly twenty years to emerge. The theory of compound radicals was generally accepted as the result of the discoveries of cacodyl (by Bunsen), and the benzoyl series (by Liebig and Wöhler); but ideas of "constitution," that is, the arrangement of atoms in the molecule, were extremely crude, and necessarily so because the facts connected with the combining capacity or valency of the elements were still unknown. The type system of classification was also undeveloped, and the work of Williamson and Frankland was wanted before further progress could be made. But before proceeding to trace their history it will be convenient to deal at once with the consequences of the application of Avogadro's principle to chemical doctrine. It was fitting that the exposition should have been undertaken by an Italian man of science, Stanislao Cannizzaro, but it was not given to the world till 1858, or nearly half a century after the publication of the essay by Avogadro, the author of this fundamental idea.

CHAPTER XIV

CANNIZZARO [1]

THE career of STANISLAO CANNIZZARO was completed in an age and country full of romance. Born as he was under the reign of a Bourbon in the kingdom of the two Sicilies, he lived to see the miserable conditions which beggared and enslaved his own compatriots swept away; he took a part as soldier and Senator in the regeneration of Italian nationality, and during the latter half of his long life he enjoyed the freedom which belongs to a united people under a constitutional monarchy.

His experiences as a man of science were no less remarkable, for it may be said he began work almost before modern chemistry, of which he helped to lay the foundations, had been called into existence. When Cannizzaro was twenty years of age, Liebig in Germany, and Dumas in France were at the height of their fame; while in England Williamson's ideas were beginning to attract serious attention. Many years had yet to elapse before a real system could be applied to the masses of facts then so rapidly accumulating.

Stanislao Cannizzaro was born in Palermo on July 13, 1826.[2] The family came from Messina, and its members at different times held important offices in that city and elsewhere in Sicily. Stanislao's father, Mariano Cannizzaro, was born in Messina, but he became a magistrate in Palermo and Minister of Police, and later President of the Gran Corte dei Conti. The mother was Anna di Benedetto, a member of a noble Sicilian house. There was a large family, of which Stanislao was the youngest. He

---

[1] The following account is adapted from the Cannizzaro Memorial Lecture given to the Chemical Society on June 6, 1912, by the author.

[2] For the facts relating to his father's life, I am indebted chiefly to Mr. Mariano Cannizzaro.

Al Prof. William A. Tilden
Roma il 13 Aprile 1898
S. Cannizzaro

[Face p. 174]

was educated partly at the Reale Collegio Calasanzio, where he won prizes, with distinction especially in mathematics. As may be imagined, the school curriculum in Sicily, as in the whole of Southern Italy, in Cannizzaro's youth was entirely under the control of the priests. Education, "frowned on as a design of the Liberals to revolutionise the State, was so successfully discouraged that in 1837 it was calculated that 2 per cent. of the rural population could read, and not very much more of the dwellers in the towns."[1] The subjects were confined to the classical languages, grammar, and rhetoric, with a little mathematics.

In 1841, at the age of fifteen, Cannizzaro began the study of medicine at the University of Palermo, and especially the study of physiology under Professor Foderà. The University was at that time in a very imperfect condition, degrees being conferred only in the faculties of medicine, law, and theology. Cannizzaro took no degree, but in 1845 proceeded to Naples, where his sister Angelina had married the Marquis Ruffo, son of King Ferdinand's Prime Minister. Here, after taking part in the proceedings of the physiological section of the scientific congress held in that year, he made the acquaintance of the famous physicist Melloni, and after working for a short time in his laboratory he proceeded, with a warm recommendation from Melloni, to Professor Piria at Pisa. The influence of Piria over his young assistant was fortunately sufficient to determine the latter to devote himself permanently to chemistry. Piria was just then at the height of his fame, justly following his discovery of the constitution of salicin, a very noteworthy feat in those early days of organic chemistry.

Cannizzaro, although an enthusiastic student, could not escape the effects of the political agitation which exercised an influence so powerful on his compatriots at that time. Those were dark days in the history of the country, and the atrocities committed in the name of order by Ferdinand's government had aroused not only the spirit of the Sicilians, but the indignation of, at least, the English people. Beside, Italian soil was occupied in the north by the armies of Austria, there was clerical misrule in the Papal States, and throughout Europe revolution was the order of the day. Cannizzaro responding to the prevalent feelings of patriotic fervour, joined in the premature rebellion in Sicily.

1 Trevelyan's *Garibaldi and the Defence of the Roman Republic*, p. 55.

Returning from Pisa to his native country in 1847, he joined the Sicilian artillery, and commanded a battery at Messina. After the fall of Messina he was sent to Taormina with a Government commission to oppose the advance of the Neapolitan troops under the General Principe Filangeri, but after March, 1849, the defeat at Novara, and the abdication of Charles Albert, the Sicilians were obliged to retreat towards Palermo, Cannizzaro being among the last to oppose the Neapolitans. On the fall of the Sicilian Government he embarked with some others on board the frigate *Independente*, which, escaping the Neapolitan fleet, succeeded in reaching Marseilles. After some months Cannizzaro made his way to Paris, and having found admission, presumably through the introduction of Piria, into the laboratory of Chevreul, he resumed his chemical studies. Here he joined Cloëz in work on cyanogen chloride and the production of cyanamide, and their results, published in 1851, constituted Cannizzaro's first contribution to the records of chemical research.

At the close of 1851 he was able to return to Italy, having been appointed professor at the National School at Alessandria, where he had the advantage of a small laboratory and the services of an assistant, "un farmacista giovane intelligente" (letter to Bertagnini). Here he was so occupied, body and mind, with his teaching that, as he complained to his friend Bertagnini, he had little hope of being able to pursue his own studies. Notwithstanding these unfavourable conditions, however, he discovered in 1853 the alcohol corresponding to benzoic acid, which he obtained by the action of potassium hydroxide on benzaldehyde, and which he continued to study during several succeeding years.

The summer holiday of 1852 was spent with Bertagnini, who had a small private laboratory at Montignoso, and here the friends carried out work on anisic alcohol, which, however, was not published until 1856. In 1854 Piria, in association with Matteucci, produced the first number of the new journal *Il Nuovo Cimento*, which was to be the organ of the Pisan school, and to the second volume Cannizzaro made the contribution referred to above. *Il Nuovo Cimento* was not established without some suspicion on the part of the Censor, the Chancellor Cardinal Archbishop, that chemistry and physics, "scienze pericolose," might cause some damage to the faith (*Nuova Antologia*, June, 1911, p. 490).

In 1855 Cannizzaro accepted an invitation to the Chair of

Chemistry in the University of Genoa, at the same time Piria[1] being transferred to Turin, while Bertagnini was appointed to replace him at Pisa.

At Genoa there was at first no laboratory, and it was only in the year following his appointment that Cannizzaro could obtain rooms in which to carry on his work.

At this time, or probably earlier, he must have begun to meditate on those fundamental questions in chemical theory which led to the famous "Sunto di un Corso di Filosofia Chimica." But his philosophical and scientific studies, as well as his teaching, were destined to be once more interrupted by the political events which at this time followed one another so rapidly in Italy. In the spring of 1860, the discontent of Southern Italy, responding to the unhappy events in the North, found vent in the insurrection which broke out in April of that year, although it was crushed almost immediately by the Neapolitan Royalist troops. However, Garibaldi with his famous thousand succeeded in effecting a landing at Marsala, in Sicily, on May 11th, and ultimately forced his way into Palermo. The story has been often told, and is full of the most astounding and romantic incidents.[2] As soon as Garibaldi had entered Palermo, Cannizzaro started for Sicily with the second expedition under General Medici, although he took no part in any battle. In Palermo he became a member of the Extraordinary Council of State of Sicily.

In October of the following year, 1861, he was called from Genoa to his native town, and was appointed Professor of Chemistry in the University of Palermo. Here, again, he had no laboratory, and it was only in 1863 that provision was made for practical work. His activity extended beyond the duties of the office he held in the University, for beside occupying a position on the Municipal Council he made great efforts to secure the establishment of schools, which were almost entirely wanting, as well as to provide for the higher education of women.[3] Later he

---

[1] That Piria was the founder of the Italian School of Chemistry was attested by Liebig. Piria held Cannizzaro in high esteem, which was repaid by the admiration of the pupil, and expressed many years later in Cannizzaro's *Vita e opere di R. Piria*, 1883.

[2] See Trevelyan's *Garibaldi and the Thousand.*

[3] *Nuova Antologia* (June, 1911, p. 492) gives a full account of his benevolent exertions in this and other directions.

became Rector of the University, and in 1867 he acted as Commissioner of Public Health during one of the severe outbreaks of cholera, in the course of which he lost a sister, struck down by the disease while nursing the sick.

Cannizzaro remained in Palermo about ten years, and during this period the work he was able to accomplish in chemical research related chiefly to the derivatives of benzylic alcohol and other aromatic substances. It is interesting to recall in this connection the fact that among the young men who came under his influence at that time was one whose name a very few years later became renowned throughout the chemical world on account of the great memoir (1874), in which was established once for all the principle by which the orientation of all the derivatives of the so-called aromatic substances can be determined. Körner's rule is familiar to even junior students of organic chemistry.

In 1871 Cannizzaro was called to Rome to occupy in the new University the Chair of Chemistry, which he retained until death summoned him away so many years later. Even in the capital city he again found no laboratory, and he was obliged to suspend his researches whilst occupied in organising the chemical institute which found shelter in the old monastic buildings in the Via Panisperna. Here he ultimately established a school, and in spite of the heavy official duties which devolved on the professor he continued during many years the study of the complex and interesting compound, santonin, and worked out its constitution with the co-operation of his pupils and assistants, Amato, Carnelutti, Gucci, Sestini, Valente, and others.

At the same time that he received the call to the University he was made a Senator of the kingdom, and as a Moderate Liberal played his part in shaping the Constitution, and establishing reform in the affairs of the now united Italy. Among other duties which fell to his lot was the organisation of the Customs laboratory and the Regia dei Tabacchi. He was also a member of the higher Council of Public Instruction, of which for some time he was President. He further occupied himself with the provision of public instruction in agriculture, and generally in helping forward the advancement of science and of the liberal professions in Italy. When the Congress of Applied Chemistry met in Rome in the year 1906, Cannizzaro was Honorary President, and it was gratifying to the visitors from so many lands to see the vivacity and energy with which the old man, then in his eightieth year, entered into all the proceedings. He was still lecturing, and some of the members had the privilege of hearing him address his class in the lecture room of the Chemical Institute. It was from this room four years later that his remains were borne by a company of his students to their last resting-place. We are informed that he continued to lecture until the year before his death: "for him to teach was to live." As soon as he perceived that his strength was failing so much that he could not lecture, all his ailments appeared to increase, and the end soon came. He died on May 10, 1910.

Cannizzaro married in Florence, in 1856 (or 1857 ?), an English lady, Henrietta Withers, daughter of the Rev. Edward Withers, who held a living in Berkshire. He left one son, who practised in Rome as an architect, and a daughter.

Active as he was as an investigator in the domain of organic chemistry, Cannizzaro's chief claim to the admiration of his contemporaries and to a distinguished place in the history of modern chemistry is based on the systematic course of theoretical teaching which he sketched in 1858.

To form a just estimate of the influence exercised on the progress of scientific chemistry by Cannizzaro's famous essay, a brief review of the state of knowledge and opinion in the chemical world up to and about the year 1858 is necessary.

The Atomic Theory of Dalton was just fifty years old, and although well rooted in the literature of chemistry, there were not a few who still refused to recognise it. Evidence of the persistence of this attitude so late as 1869 is afforded by Williamson's lecture,[1] and especially by the discussion which ensued upon it. Some thought to perceive a distinction between physical atoms and chemical atoms, but generally they seem to have retained the fundamental notion of Dalton, which conceives each atom to be a sphere existing either alone or in close contiguity with other similar atoms, and separable more or less from one another by the influence of heat. Students at this time were generally unfamiliar with the word "molecule,"[2] for chemists spoke as complacently, and in a sense as justly, about an *atom*

---

[1] *Journ. Chem. Soc.*, 1869, **22**, p. 328.

[2] The word *molecule* was occasionally used by Dalton, *e.g.*, *Chemical Philosophy*, vol. i., p. 70, and in the sense of Atom by Ampère (*Ann. Chim. Phys.*, 1814, **90**, p. 43).

of water as about an atom of oxygen. For the most part, also, they had never heard the name of Avogadro. Considerable advances had been made toward the estimation with exactitude of what were then usually, although incorrectly, called "atomic weights," notably by Berzelius, Dumas, Pelouze, de Marignac, and Stas. The figures thus afforded by experiment were only equivalents or combining proportions, uncorrected by reference to any standard, for the excellent reason that there was no standard generally recognised; and even in the use of the term "equivalent" there was the utmost confusion, of which evidence is provided by the statement in one of the most widely circulated text-books of the period (Fownes, 1856) that the numbers called equivalents "represent quantities capable of exactly replacing each other in combination," the list of numbers referred to including nitrogen 14, carbon 6, whilst hydrogen was 1, and all were said to be equivalent to oxygen taken as 8.

The consequences of bringing equivalents to the same volume were at this time, and even much later, not considered by the great majority of teachers, and although vapour densities were frequently the subject of experiment, the results were used merely to check the empirical formula deduced from analysis of the substance, and few thought of adopting a standard volume and revising the empirical formula so as to harmonise with it. If, for example, the vapour density of acetone was found, it would be used merely to substantiate the formula deduced from analysis, namely, C_3H_3O ($C = 6$, $O = 8$), and "whether the rational formula of acetone is C_3H_3O or $C_6H_6O_2$ or $C_9H_9O_3$ the vapour density does not enable us to decide" (Galloway's *Second Step*, 1864, p. 68).

This is surprising in view of the fact that so far back as 1826 Dumas, in the memoir in which he describes his well-known method of taking vapour densities,[1] refers to the fact that all physicists agree in supposing that in elastic fluids under the same conditions the molecules are placed at equal distances, or in equal numbers in the same volume.

Up to this time also the conception that the ultimate particles of the elements themselves might contain more than one atom had not been commonly accepted. It was believed that

[1] "Sur quelques points de la Théorie Atomistique" (*Ann. Chim. Phys.*, 1826, **33**, p. 337).

combination could only occur between substances of opposite chemical or electro-chemical character, hydrogen with oxygen, for instance; but that hydrogen could unite with hydrogen, or oxygen with oxygen, was not generally admitted.

It is evidence of the complete neglect with which Avogadro's great memoir of 1811 had been treated that chemists generally at this time did not know, or had completely forgotten, that the constitution of elementary molecules in the gaseous state had been very clearly explained by him. In the second division of his memoir he discusses the case of the "elementary molecules," and his position is indicated clearly enough by the case of water. He says: "Ainsi la molécule intégrante de l'eau, par exemple, sera composée d'une demi-molécule d'oxigène avec une molécule, ou, ce qui est la même chose, deux demi-molécules d'hydrogène." This view of the constitution of elementary molecules has already been explained (p. 172).

Gerhardt in 1843 was led to represent elementary hydrogen as hydrogen hydride, HH, and gaseous chlorine as chlorine chloride, ClCl, but this conception of the constitution of elementary molecules was not derived from any direct consideration of the views of Avogadro or Ampère, whose names are not mentioned.

Gerhardt's system of formulæ appears to have been based chiefly on his own view that every chemical change is a form of double decomposition, and he is at great pains to show that in chemical reactions, whether of combination or decomposition, the proportion of water or of carbonic acid involved was never less than the amount represented by the formulæ H_2O and CO_2, in which $H = 1$, $O = 16$, and $C = 12$; and, similarly, the amount of free oxygen or hydrogen was never less than the amount represented by O_2 and H_2 with the values as just stated.

During all this long period the name of Avogadro had been treated with a neglect which is scarcely compensated by the recognition now accorded to it nearly a century after his time. Among the French, Ampère got more credit in this connection than seems to belong to him, for his paper (*Ann. Chim. Phys.*, 1814, **90**, p. 43) three years later than the memoir of Avogadro shows little evidence that he attached the same importance to the theorem that equal volumes of different gases under the same conditions contain the same number of particles as did Avogadro. The memoir of Ampère is chiefly devoted to a consideration of the probable forms of the "particles" (molecules) of crystallised substances.

Even those chemists who are generally supposed to have made use of Avogadro's idea have neglected all reference to its origin. Gerhardt, for example, ignored the Italian physicist, and Dumas in the paper on vapour densities already quoted only mentions his name in the following passage, which forms the conclusion of the memoir: "Nous sommes bien éloignés encore d' l'époque où la chimie moléculaire pourra se diriger par des règles certaines, malgré les avantages immenses que cette partie de la philosophie naturelle a retirés des travaux de MM. Gay-Lussac, Berzelius, Dulong et Petit, Mitscherlich, ainsi des vues théoriques de MM. Ampère et Avogadro. L'activité singulière de M. Berzelius et le bon esprit des chimistes dont il a enrichi l'Allemagne pourraient cependant faire espérer sur ce sujet important une révolution prochaine et durable."

It was not, however, until thirty years later that this revolution was brought about, and its author was a chemist from no northern school. The year 1858 must for ever be distinguished in the history of chemistry, for it was then that Cannizzaro led the way out of the darkness in which all had been so long struggling.

After this preamble we may more easily realise the nature and extent of the revelation, as it may well be called, which students of chemistry owe to Cannizzaro. That it remained for some years almost unknown may be attributed in part to the barrier constituted by the language in which his essay was originally published. But it is not creditable to the chemists of 1860 that the Congress held at Carlsruhe in September of that year, at which Cannizzaro was present and expounded his views, should have dispersed without a general acceptance of the fundamental principles which to us seem unassailable. The only excuse which presents itself now is the fact that at this period the difficulties arising out of dissociation of compounds like sal-ammoniac and sulphuric acid when volatilised by heat, and which gave rise to the so-called anomalous vapour densities, had not been cleared away. To contend, as some speakers seem to have done, that these subjects are matters of opinion, and that every scientific man is entitled to perfect freedom in respect to the views he adopts, is to misunderstand the case. In art, in which field sentiment, emotion, and taste are the only considerations involved, complete freedom is clearly necessary, but in science whenever facts have been established and an agreement has been arrived at in regard to fundamental assumptions, reason ought to be the only, as it is the sufficient, guide. Unfortunately, this has not always been the case.

It is only fair to mention that of those chemists who were present at the Carlsruhe Congress in 1860, one at least came away convinced. In a prefatory note to the German edition (published in 1891) of Cannizzaro's *Sketch,* Professor Lothar Meyer relates how he received at the meeting a copy of this paper, which he read with surprise at the clearness with which all the most important difficulties were removed. He says: "It was as though scales fell from my eyes, doubt vanished, and was replaced by a feeling of peaceful certainty." In 1864 Meyer published his well-known treatise on the *Modern Theories of Chemistry,* in which the views of Cannizzaro are fully developed.

To those who have read Cannizzaro's *Sketch of a Course of Chemical Philosophy,* of which a belated French translation has been produced by the Alembic Club, it must be a matter of wonder that the facts and arguments set forth should not have been sufficient to have cleared away the previous confusion immediately. With small and unimportant corrections, it represents a course of instruction which might have been given as embodying the accepted views of the chemical world down to quite recent times, and a perusal of this essay, even now, would be of the utmost value to many teachers.

Cannizzaro's *Sketch* begins with the following words: [1] "I believe that the progress of science made in these last years has confirmed the hypothesis of Avogadro, of Ampère, and of Dumas on the similar constitution of substances in the gaseous state; that is, that equal volumes of these substances, whether simple or compound, contain an equal number of molecules; not, however, an equal number of atoms, since the molecules of the different substances, or those of the same substance in its different states, may contain a different number of atoms, whether of the same or of diverse nature."

The author then proceeds to trace the history of this conception, of the consequences to chemical theory, and of the ideas which prevented the immediate acceptance of this hypothesis,

[1] I have made use of the *Alembic Club* version in these quotations.

and the confusion which resulted from the failure to distinguish molecules from atoms. In order to bring harmony into the various branches of chemistry, he then shows that by applying the hypothesis of Avogadro the weights of molecules may be determined before their composition is known, and that a knowledge of their composition is not necessary to this end. Having settled the molecular weights of a series of substances containing one element in common, the discovery is made that the different quantities of the same element contained in different molecules are always whole multiples of one and the same quantity, which represents the atomic weight. After studying the constitution of various volatile chlorides, bromides, and iodides, the question of the constitution of mercuric and mercurous compounds arises, and the author proceeds to show that the smallest proportion of mercury present in any molecule containing that element is 200, and that this is therefore the atomic weight of the metal. This number is then confirmed by appeal to the law of specific heats. The analogy of the chlorides of copper with those of mercury next leads to the examination of these compounds, but as the vapour densities of these salts are not known, the specific heat of copper and of its compounds leads to the number 63 as the atomic weight of copper. Whether this is the molecular weight of the uncombined metal there is no means of knowing until the vapour density of this substance can be determined. Many other metals are then examined, and the author points out that in such cases as tin, which produces compounds volatile without decomposition, and of which the molecular weight can be determined, the atomic weight deduced from specific heat is in agreement with that deduced from vapour density. But then the question arises: "Are the atoms of all these metals equal to their molecules, or to a simple submultiple of them?" And he proceeds: "I gave you above the reasons which make me think it probable that the molecules of these metals are similar to that of mercury; but I warn you now that I do not believe my reasons to be of such value as to lead to that certainty which their vapour densities would give if we only knew them." Herein he differs from Gerhardt, who had represented the atoms of all the metals as fractions of the respective molecules, as in the case of hydrogen.

A little later Cannizzaro came very near to the modern idea of valency when discussing the capacity of saturation of different atoms. When referring to diatomic radicals as "those which, not being divisible, are equivalent to two of hydrogen or to two of chlorine," he proceeds to show "that cacodyle, C_2H_6As, methyl, CH_3, ethyl, C_2H_5, and the other homologous and isologous radicals are like the atom of hydrogen, monatomic, and, like it, cannot form a molecule alone, but must associate themselves with another monatomic radical, simple or compound, whether of the same or of a different kind, and that ethylene, C_2H_4, propylene, C_3H_6, are diatomic radicals analogous to the radicals of mercuric and cupric salts, and to those of the salts of zinc, lead, calcium, magnesium, etc.; and that these radicals, like the atom of mercury, can form a molecule by themselves. The analogy between the mercuric salts and those of ethylene and propylene has not been noted, so far as I know, by any other chemist."

There is much more in the *Sketch* which was important for the elucidation of the views put forward by the author, but the extracts given are sufficient to show how clear, how systematic, and how logical was the mind which could thus choose out from the tangled mass of fact and fiction constituting chemical theory in his day, the materials for a consistent, orderly, and productive system of scientific chemistry.

What Cannizzaro did for chemistry may be broadly stated under the two following heads:

First, he laid down for all time the two principal methods by which atomic weights are determined, the one by reference to the molecular weights derived from an application of Avogadro's rule, and the other by the adoption of the principle originally discovered by Dulong and Petit as to the general relation of atomic weight to specific heat among the solid elements, and he showed that these two methods when applicable to the same case lead to the same results.

Secondly, he placed inorganic chemistry in a new light by applying to inorganic compounds the same principles which had been applied to organic compounds, and thus finally disposed of the superstition which had hovered so long in the minds of chemists that organic chemistry was subject to laws different from those prevailing among mineral substances.

There is, in fact, but one science of chemistry and one set of atomic weights.

It will not be without interest to recall some of the consequences of the ultimate adoption, tardy as it was, of the principles laid down by Cannizzaro. The unanimity which has prevailed among chemists during the last forty years or more as to the fundamental principles inculcated by Cannizzaro is a proof that his system is not only reasonable but is practically convenient. We are not now divided into parties on the subject of atomic weights, and although some may still incline to use hydrogen as the unit, whilst others prefer an exact integer for oxygen, these differences do not affect the notation or the common language of chemistry. As a result of a uniform standard for atomic weights we now possess a natural system of classification of the known elements in the form of the periodic scheme with all its consequences, which will be described later. Out of the revised and uniform system of atomic weights we also have a universally acknowledged system of constitutional formulæ, based on valency, which we may define as the habit in regard to combination exhibited by the several elementary atoms, without necessarily forming any hypothesis as to the cause or nature of chemical "affinity." The wonderful discoveries which have been brought to light in the department of stereochemistry provide a body of evidence in favour of atomic structure which can never be set aside ; and the day is now gone by when serious support can be found for any form of anti-atomic doctrine, since we have been shown how single atoms can be seen and counted.

That all this knowledge would have come into the possession of mankind sooner or later cannot be doubted, but that this generation enjoys all the fruits of experiment in chemistry we owe to Cannizzaro. Without the clear light which his doctrine cast into the dark places of chemical theory sixty years ago chemistry might have remained a mass of unclassified, incoherent, and perplexing facts.

This is why it is incumbent on this generation of chemists to do honour to his memory. The English Societies cannot be charged with indifference to the great services rendered to science by Cannizzaro, for, as mentioned at the outset, his name was placed on the limited roll of Honorary Members of the Chemical Society so far back as 1862. Ten years later he was invited to give the second Faraday Lecture, and again in 1896, on the occasion of his seventieth birthday, an address was presented to him on behalf of the Chemical Society, in which full expression was given to the feelings of respect and admiration entertained by all the Fellows of the Society toward the veteran chemist.

The Royal Society, also, awarded to him in 1891 the Copley Medal, which is regarded as the highest honour in the power of the Society to bestow.

The case of Avogadro and that of his great exponent Cannizzaro suggest that in remembrance of the unity of scientific thought throughout the world, fed by the contributions of all nations, a plea may be entered, not only for justice to individuals, but for complete international impartiality in matters of science.

Italian science is no mushroom growth. Before our own Royal Society was founded, or perhaps even thought of, before the French Academy of Sciences came into existence, Galileo and Torricelli were making discoveries of world-shaking significance. In those times, however, to have observed natural phenomena, and even to be suspected of holding unfamiliar, novel, and therefore heretical opinions about the world in which man is placed, was to draw down on the unhappy philosopher the condemnation of political and ecclesiastical ignorance and fanaticism. No wonder that those whose interest was excited by the new knowledge then coming to light endeavoured to conceal their discussions and places of meeting under all kinds of fantastic and often ridiculous masquerade.[1]

Happily such prejudice, although occasionally showing itself, as in the instance already mentioned of the publication of *Il Nuovo Cimento*, is now powerless, and the neglect of Avogadro's hypothesis cannot be put down to the influence of ecclesiastical authority. The obscurity which prevented its recognition arose out of the very nature of chemistry itself, and even the prosecution of research seemed for a time only to add to the prevailing confusion by producing crowds of new and unclassified facts. Science had, in fact, to wander in the wilderness until the great leader came to show the way. That Avogadro's life should have come to a close only two years before the formal proclamation and application of his doctrine before a congress of chemists seems a harsh dispensation, but if he had lived only a little longer it would surely have been to him an added satisfaction that that doctrine should have been established by his fellow-countryman.

---

[1] See, for example, Disraeli's *Curiosities of Literature:* "On the Ridiculous Titles assumed by the Italian Academies."

Group VIII

ATTEMPTS AT CLASSIFICATION

LIEBIG (1803-1873) DUMAS (1800-1884)

CHAPTER XV

LIEBIG [1]

LITTLE more than a hundred years ago Europe was plunged in the misery of war. We who have survived the horrors of the recent war can appreciate the condition of the world at that time. Then, as in our day, almost every country had suffered the bitter experience of seeing the devastation caused by the passage of contending armies, the death and suffering of thousands of fighting men, and the want and desolation spread over still greater numbers of a helpless population. Amid all the wretchedness of the time, insecurity of property, dearness of food, frequent changes of governments, and every condition which would appear to be unfavourable, the study of nature steadily went on. France, still staggering from the shocks of the revolutionary period, had still many distinguished men of science, Laplace, Berthollet, Lamarck, Cuvier, while the memory of Lavoisier was fresh and green, and Gay-Lussac, Dulong, Arago, and Chevreul were among the coming men. England, still engaged in the struggle with Napoleon, possessed Humphry Davy, Rumford, and Dalton, and Herschel among the astronomers. Henry Cavendish was still living, though an old man, and Priestley was but lately dead. In Germany, Goethe might be counted among the votaries of science, and Prussia had sent forth Humboldt to survey the world, while in Italy, Volta was busy in the study of electricity, and Avogadro, little noticed by the world, was meditating on the properties of gases and preparing for the enunciation of the great principle which is now associated with his name, though it took the chemical world half a century to recognise it. Berzelius, then young, was preparing, by his eager

---

[1] The following account is adapted from a lecture delivered at Oxford on August 23, 1911, at the Fifteenth Summer Meeting, by the author.

[*Face p.* 188]

activity in research, for that great position of almost undisputed authority in the chemical world, which, as already shown, he filled for nearly forty years.

Born in Darmstadt, on May 12, 1803, where his father was a colour manufacturer, Justus Liebig passed through an unsuccessful school career at the local gymnasium, and, at the age of sixteen, was apprenticed to an apothecary. It soon became evident, however, that he was as little fitted to become a pill-maker as he was to be a Greek scholar, and he ultimately persuaded his father to allow him to go to the then newly-founded University of Bonn, whence he followed Kastner, the professor of chemistry, to Erlangen. But Liebig soon became convinced that he could not study chemistry effectively in Germany, and after taking his degree at Erlangen, at the age of nineteen, he proceeded to Paris. There, after many difficulties, he ultimately obtained the privilege of working in Gay-Lussac's laboratory, where he remained about two years. In 1824, on the recommendation of Humboldt, he was appointed Extraordinary Professor of chemistry at Giessen, being then only twenty-one years of age. He became Ordinary Professor two years later, and remained at Giessen until called to Munich, in 1852. There he died on April 18, 1873.

Such was the main course of Liebig's career; but to draw a picture of the man from descriptions of his personal characteristics is not so easy. In early youth he became familiar with the poet Platen, who noted in his diary "the friendly earnestness in his regular features, great brown eyes, with dark shady eyebrows, which attracted one instantly."

Those brown eyes, shining with earnestness, remain in the portraits which have come down to us, and, as a family feature, reappear in the faces of some of his children. Ardent, eager, enthusiastic in the pursuit of experiment, his remarkable power of exact observation stood him in good stead. Kindly and tender with children, there were times when eagerness in research or controversy led to exhibitions of impatience, but the steadfast character of the man is illustrated by the persistence of his lifelong intimacy with Friedrich Wöhler. This intimacy resulted in a correspondence which extended over more than forty years, and had consequences in the lives of both men which were full of importance for the progress of chemical science. To this reference must be made further on.

We may now endeavour to sketch, in outline, the state of knowledge and theory when Liebig entered on his career.

The modern use of the term element, which had been introduced by Boyle in the 17th, century, was by this time universally adopted, and to the list which included the metals had been added such important substances as oxygen, hydrogen, nitrogen, and chlorine. To use the words of Davy, in one of his researches on chlorine, "Neither oxygen, chlorine nor fluorine are asserted to be elements; it is only asserted that they have not been decomposed." And that is the sense in which the term in modern times has always been used. The process of burning or combustion was, of course, now always explained by Lavoisier's doctrine, according to which a body in burning combines with the oxygen of the air, and forms one or more chemical compounds with it. At the time that Liebig went to Giessen, in 1824, Sir Humphry Davy was still living, but his scientific career was practically closed, and Berzelius was the predominant authority in matters of theory. Gay-Lussac, in Paris, had made important discoveries relating to the proportions in which gases enter into combination. Dalton's atomic theory, propounded in 1808, though not generally accepted, was gaining ground. Broadly, the position was this: elements were clearly distinguished from compounds, chemical combination was explained by the supposition that it was due to the close approximation of atoms of opposite kinds, and the union of atoms to form a chemical compound was attributed to the attraction caused by charges of electricity of opposite nature, which were supposed to be resident on the atoms.

But the composition and nature of "organic" compounds were practically unknown. A few such substances had been isolated, *e.g.*, milk sugar and grape sugar were known as distinct substances, and were differentiated from common sugar. Alcohol, nearly pure, had been known, in the form of spirit of wine, from early times. Acetic acid was known, as well as several acids found in vegetable tissues, such as oxalic, formic, malic, tartaric, and benzoic acids. There were, however, no means of determining their composition, and although Lavoisier had devised an apparatus in which organic compounds could be burned in oxygen, and the water and carbon dioxide thus formed could be collected, the process was both cumbrous and incapable of yielding exact results.

A most interesting autobiographical sketch was discovered among Liebig's papers many years after his death, and from this we learn that in his early life "at most of the universities there was no special chair for chemistry. It was generally handed over to the professor of medicine, who taught as much as he knew of it, and that was little enough, along with toxicology, materia medica, etc." But the total neglect of experiment was the source of much mischief, and the persistence of the degenerate deductive method led to neglect of the careful observation of nature. Liebig describes the lectures of Prof. Kastner as without order, illogical, and resembling the jumble of knowledge which he carried about in his own head. When he got to Paris all was different, and the lectures of Gay-Lussac, Thénard, and Dulong had for the young student an indescribable charm. The lecture consisted of a judicious series of demonstrations—experiments of which the connection with each other was pointed out and explained ; and soon the consciousness dawned on him that all chemical phenomena, whether exhibited by the animal, vegetable or mineral kingdoms, are connected together by fixed laws.

Liebig therefore returned from Paris to his own country with the intention of founding an institution in which students could be instructed in the art and practice of chemistry, the use of apparatus, and the methods of chemical analysis. In view of the total absence of such provision elsewhere, it is not surprising to learn that, so soon as its existence became known, students streamed into the Giessen laboratory from every civilised country. It is interesting to learn from Liebig's own words what was the method he adopted. Obviously, in order to teach a large number at one time, it is necessary to have a systematic plan, and in his case this had first to be thought out and then put to the proof, as no course existed which could be used as a model. He says, however, that "actual teaching in the laboratory, of which practised assistants took charge, was only for the beginners ; the progress of my special students depended on themselves. I gave the task and supervised its carrying out. There was no actual instruction. Every morning I received from each individual a report on what he had done the previous day, as well as his views about what he was engaged on. I approved or criticised. Every one was obliged to follow his own course. In the association and constant intercourse with each other, and by each participating in the work of all, everyone learned from the others. Twice a week in winter I gave a sort of review of the more important questions of the day. We worked from break of day till nightfall. Dissipation and amusements were not to be had at Giessen. The only complaint which was continually repeated was that of the attendant, who could not get the workers out of the laboratory in the evening when he wanted to clean it."

Such was the spirit and such the method by which a great school was created! Nor was this the only result. To the influence and example of the school at Giessen may be attributed the rapid spread of the new method of teaching chemistry. In 1824 there were no laboratories devoted to the purposes of instruction. A few of the most eminent professors of chemistry —Berzelius in Stockholm, Gay-Lussac in Paris, for example—admitted one or two students already advanced in the subject to practise in their private laboratories, but only as a great favour. In this way Mitscherlich, Rose, Wöhler, and Magnus had repaired to Berzelius in Stockholm as Liebig had gone to Paris. But in a few years the fame of what Liebig was doing in Giessen penetrated to other countries of Europe, and many of the men who had studied under his direction became teachers in other lands. Here in England no chemical laboratory for general instruction existed, and only in the medical schools were a few tests described and shown. In London the Society of Apothecaries had a laboratory which had existed since 1671; but this was used not for teaching, but as a place of manufacture of drugs for use in medicine. At Cambridge the professor of chemistry was a country clergyman, who came up once a year to give a course of lectures. At Oxford the professor of chemistry was also, later, professor of botany, and in neither university was there a laboratory for instruction, nor was chemistry a subject recognised in the curriculum for a degree. Twenty years later things began to improve. In this country the first laboratory for instruction in practical chemistry was provided by the then newly instituted Pharmaceutical Society of Great Britain at their premises in Bloomsbury Square. This was in 1844, and in the following year a new and enlarged laboratory was fitted with places for twenty-one students.[1] About this time the Royal College of Chemistry was

[1] I saw this laboratory about 1857. It had the aspect which one usually associates with ideas of the alchemists. Many of the operations were connected with the use of furnaces, such as fusion, sublimation, etc., and the place was full of smoke and fumes.

Leben und Treiben im ersten chemischen Laboratorium Justus Liebigs in Gießen. Die Laboranten im Hauptarbeitssaal. Nach der Zeichnung von Trautschold 1842.

No. 1. Ortigosa (Mexikaner). No. 2 und 3. Namen unbekannt. No. 4. Keller. No. 5. Dr. Will, Assistent Liebigs, später Professor der Chemie und Nachfolger Liebigs in Gießen. No. 6. Adolf Strecker, Assistent Liebigs, 1860 Professor der Chemie in Tübingen und 1870 in Würzburg. No. 7. Aubel, Famulus. No. 8. Wydler aus Aarau. No. 9. Varrentrapp. No. 10. Scherer, Professor. No. 11. Namen unbekannt. No. 12. Emil Böckmann. No. 13. A. W. Hofmann, Assistent Liebigs bis 1845, später Professor der Chemie in Berlin, Begründer der Deutschen Chemischen Gesellschaft.

established in temporary quarters in George Street, Hanover Square, and soon afterwards the Birkbeck Laboratory, modelled on that of the Pharmaceutical Society, was built at University College. Many other laboratories were opened about this time. In 1848 Pélouze founded in Paris a laboratory to which some English chemists resorted. But the Giessen laboratory under Liebig's direction continued to supply the majority of the teachers who in the succeeding generation founded schools, not only in Germany, but in other countries—Hofmann, for example, at the Royal College of Chemistry, and Williamson, who was appointed at University College in 1849.

Liebig's career as a chemist and investigator was influenced in no small degree by his friendship with Wöhler. Born three years before Liebig, Friedrich Wöhler studied medicine at Marburg, but subsequently pursued chemistry at Heidelberg under Leopold Gmelin. Having relinquished medicine on taking his degree, he obtained the privilege of working with Berzelius in his laboratory at Stockholm. On his return from Sweden in 1824 he was appointed teacher of chemistry in the Trade School in Berlin. Some years later he became professor in the University of Göttingen. Soon after his return from Sweden he met Liebig in Frankfort, and a close intimacy at once sprang up, which continued for more than forty years, to the end of Liebig's life. Two volumes of their correspondence have been compiled by Hofmann, and the perusal of these letters, extending from 1829 to 1873, affords a view of the subjects which occupied the minds of both, as well as many of the incidents of their lives. Liebig paid several visits to England, and in a letter to Wöhler dated from Giessen, November 23, 1837, he tells him that he has travelled through England, Ireland and Scotland in every direction, and has seen many surprising things, but has learned little. The absence of scientific knowledge in England he attributes to the badness of the teaching. In another letter, addressed to Berzelius nearly at the same time (November 26th), he says: "England ist nicht das Land der Wissenschaft," [1] only there is a widespread "dilettantismus," and he complains that "die Chemiker schämen sich Chemiker zu heissen, weil die Apotheker, welche verachtet sind, diesen Namen an sich gezogen haben." [2]

[1] "England is not the land or home of science."

[2] "The chemists are ashamed to call themselves chemists, because the apothecaries, who are despised, have appropriated the name."

Liebig's contributions to pure chemistry, though so numerous and important, can be recalled only briefly. They may be placed under three heads, namely, first, the invention and perfecting of a method for analysing organic compounds, which in all essential features is still practised everywhere.

Secondly, the discovery of a large number of new compounds, of which even the names cannot now be mentioned for want of space, but which include chloroform and chloral and many cyanides. He also established the formula of uric acid and the nature of aldehyde.

Thirdly, we owe to Liebig the conception of the theory of compound radicals, which arose out of his researches jointly with Wöhler (1832) into the products from essential oil of bitter almonds.

In a letter to Wöhler (May 26, 1839), Liebig writes that he is occupied with the study of the phenomena of fermentation and putrefaction, and having sent an account of his views to Wöhler, another letter, dated June 3rd, discusses the criticism which he has received from him. In the postscript to this long and interesting letter, we find a concise statement of Liebig's hypothesis concerning the action of ferments.

Liebig's explanation of these changes was based on purely mechanical ideas as to the motions of the hypothetical particles or atoms. He imagined the atoms of a substance which causes fermentation or putrefaction to be in a state of unceasing vibratory motion, and that this state of agitation was communicated to the molecules of the sugar, causing them to undergo an internal rearrangement, and to break down into simpler structures of a more stable nature, in the case of alcoholic fermentation of sugar, in fact, into alcohol and carbon dioxide.

Liebig made the mistake of ignoring, as nearly all chemists and biologists of that time ignored, the constitution of the ferment. In 1859 and following years, Pasteur, the great French chemist, demonstrated the essentially vitalistic character of the phenomenon, and showed that the destruction of the sugar was an effect concomitant with the growth and multiplication of the cells of a minute organism, visible under the microscope. A special form and character of organism is concerned in each type of fermentation.

The organised character of yeast had been proved many years before by the observations of Kützing, Cagniard Latour and

Schwann. Nevertheless, the views of Liebig prevailed for some time. In the English version of his famous letters on chemistry, in the fourth edition, which appeared in 1859, there is a chapter headed "Theory which ascribes fermentation to fungi refuted." As a matter of fact, it was about this time established.

Liebig was ultimately convinced of the organic nature of yeast, but he still contended for his theory of molecular destruction by communicated agitation, as furnishing the explanation of the physiological act which comes about within the cells of the yeast. An important step was taken much later, when, in 1897, it was shown by Buchner that something can be dissolved out of yeast which, independently of the cells, is capable of resolving sugar into alcohol and carbon dioxide. Thereupon, it seemed to some that Liebig's views might be resuscitated. But the changes which occur are now known to be very complicated, involving, in the first place, a process, not of destruction, but of building up molecules of a more complex nature, before they are broken down into the final products of fermentation. Liebig's theory, therefore, disappears from the scene.

Before 1840 it may be stated as almost literally true that physiology in the modern sense of the term did not exist, and certainly there was but a small basis for chemical physiology. The chemical production of urea independently of animal life, by Wöhler, in 1828, was a fact of which the deep significance appeared only much later. The studies in organic chemistry, into which Liebig had plunged alone, or in conjunction with his friend, necessarily attracted his attention to problems connected with the phenomena of animal and vegetable life. His visit to England, in 1837, was largely occupied with observation of the methods of agriculture then prevalent, and during the succeeding years we find in the catalogue of his scientific papers many signs of his activity in pursuit of questions connected with the application of chemistry to agriculture, the growth and nutrition of plants, the formation of fat in the animal body, the composition and classification of foods, the source of animal heat, and the chemical processes connected with respiration and digestion. It is not possible to enter freely into the discussion of all these great subjects, but a glance may be given at Liebig's views in regard to two of them, not because those views have retained their predominance, but because of the stimulus they gave to inquiry and the encouragement he gave by precept and example to the fundamental principle on which the greater part of modern science is built, namely, the constant appeal to nature, not only by observation, but by systematic experiment.

In Liebig's time all biological processes were supposed to be controlled by what was called "vital force," that is, something which is not mechanical force, nor heat, light, electricity, nor chemical affinity. We are still a long way from knowing what life is, but to show how far some physiologists have travelled in the opposite direction, a very short quotation from a recent book may be given. Concerning the use of the term "metabolism," which is a comprehensive word covering all chemical changes which go on in the body during life, the writer directs attention to its implication "that all the phenomena of life are, at bottom, chemical reactions. When a muscle twitches no less than when a gland secretes, it is not too much to say that when we are moved to tears or laughter, it is chemical reactions that are the underlying causes to which ultimate analysis must lead us." This must be regarded as an extreme view.

Turning first to Liebig's classification of foodstuffs. He endeavoured to account for the maintenance of the animal functions, the growth and repair of the body, beside the maintenance of its temperature.

Liebig attributed, as we believe correctly, the heat produced in the body to the process of burning, which goes on in the tissues in consequence of the absorption of atmospheric oxygen. Liebig was also right in his assertion that animals do not necessarily derive fat from their food, but the animal body is a laboratory, in which fat may be manufactured from carbohydrates, such as sugar and starch. The substance burned in the body is material derived from the food, but it has long been known that the substance thus burned does not consist exclusively of sugar, starch and fat, which Liebig called *respiratory* foods.

The other constituents of food, now included under the general term protein, which contain nitrogen, and are more or less like white of egg in properties, he called *plastic* foods. These were supposed to produce new tissue, or repair waste, and to be the source of muscular energy or power to do work.

It is now known that the case is by no means so simple, and, in fact, this classification now possesses only historical interest. The whole question when considered in the light of modern knowledge is, in fact, a mass of difficulties, and very far from

ıg clear of serious controversy. Liebig's name is associated
he public mind almost exclusively with the extract of meat
ch he prepared for the first time in connection with his
lies of food. This is to do him less than justice. Liebig
er proposed it for use as a substitute for meat, because it
tains only a part of the constituents of flesh. It appears that
idea, in the first instance, was to turn to account the flesh,
ch would otherwise be wasted, of animals which in Australia
South America were then bred solely for the sake of their wool
fat. Extract of meat is to be regarded as a valuable stimu-
; to be consumed together with bread or other vegetable food.

Turning to the investigations into the operations and theories
griculture with which Liebig's name is associated, the ques-
ıs arise: Whence do plants get their carbon and nitrogen,
ch, together with hydrogen and oxygen and water, form the
erial of their tissues? What is the use of the mineral sub-
ıces found in the ash left on burning vegetable matter? Why
different soils adapted to different crops, and what is it that
es fertility to a soil?

The state of knowledge on such subjects is indicated roughly
the summary which had been provided by the lectures of Sir
mphry Davy in 1813. During the subsequent twenty-five
rs very little had been done in the way of experiment, but it
ıld be only fair to mention the name of the great French
icultural chemist Boussingault as one of the pioneers a little
advance of Liebig in the study of such questions. Briefly,
position was somewhat as follows: it was known that plants
ompose the carbonic acid of the air, using the carbon and
ing the oxygen go free, but it was commonly supposed that
brown or black substance in the soil, which is usually called
nus, and is the result of the decay of preceding vegetable
wth, was the chief source of the carbon in growing vegetables.
big pointed out that this was impossible, because it failed to
w from what source the original plants from the decay of
ch humus was formed derived their carbon. Liebig was the
t to study carefully the mineral constituents of plants and to
gnise the importance of certain substances, especially potash
l phosphates. The services which Liebig rendered to the
ld in connection with plant physiology and agriculture are,
vever, less to be recognised in the shape of positive contribu-
ıs to knowledge than in the example set and in the influence

of that example in stimulating systematic investigation of agricultural questions. By 1840 Liebig was one of the most famous chemists in the world, and the effect of his inquiries is shown in the activity which became manifest almost immediately after the communication of his first report to the British Association at the Glasgow meeting in 1840. In Germany the Government instituted a large number of Versuchs Stationen in different parts of the country, and in 1843 the systematic experiments were started at Rothamsted which must for ever place the names of Lawes and Gilbert among the benefactors of the world.

Liebig died in 1873; but the period of his greatest activity in science lies further back by thirty years. Vast changes have since been brought about by chemical discovery, which, be it always remembered, is based on experimental work in the laboratory. That is the reflection which supplies the explanation of Liebig's great influence on the progress of science. That influence was fully recognised by the generation of chemists now just passed away, and it seems to be a duty to preserve as long as possible a memory so rich in past benefits and so full of suggestion for future use.

Liebig made many discoveries in chemistry; but his great and permanent service to the world was not in the isolation and study of individual compounds or series of compounds, nor in the conception of theories of chemical action, nor even in views which he promulgated concerning the operations of agriculture, the composition of food, the processes of digestion, or the source of animal heat. His great service consisted in showing how chemistry should be studied and how it should be taught, in setting the example of submitting all questions to the light obtained by direct experimental study of nature, and in thus affirming and illustrating the principle that what is called pure science is of greater permanent value than what is called applied science; a knowledge of the laws of nature is more useful than many inventions.

In the Giessen laboratory were trained a considerable number of chemists, many of whom became the teachers of the next generation. From these teachers and their pupils, guided by the same principles as those of the Giessen school, came discoveries of first-rate importance. If Hofmann, a student of Liebig's, had not been attracted to the study of aniline, an inconsiderable constituent of coal tar, if his pupil, Perkin, had not been led to a further study of its transformations, we should have had to wait a long time for the coal-tar dyes and the industries connected therewith. If a host of workers trained in Liebig's laboratory, and others emulating their example, had not cultivated the study of all sorts of carbon compounds, often unimportant in themselves, we should not have seen the numerous applications of chemistry to medicine—the saccharin, aspirin, antipyrin, sulphonal, etc.—nor the artificial perfumes, such as those of violet and lilac, which are now made independently of the original source in the flowers. Without the foundation work just mentioned we could not now have the beginnings of the true physiology based on the study of chemical and physical processes and reactions, nor the possibility of following the changes brought about by all sorts of ferments, on the combined results of which we may hope to have a complete development of a scientific system of medicine and the treatment of disease.

But there is one other direction of Liebig's activity not yet referred to. Discoveries in the study of nature are of little value unless they can be communicated to that part of the world which can and will make use of them. Up to the end of the eighteenth century there were no means of publication except, on one hand, through the transactions of the half-dozen academies, and these were the only scientific periodicals; or, on the other, by the special treatises prepared by investigators for the purpose of making known their own discoveries or opinions. Thus we have the famous works of Robert Boyle on the Spring of the Air, and the Sceptical Chemist, Scheele's works on Air and Fire, Priestley's Experiments and Observations on different kinds of Air, Dalton's New Chemical Philosophy, and many others. The publication of such books was often accomplished only after years of preparation. In 1832 Liebig founded the *Annalen*, which have ever since borne his name. Out of Trommsdorff's old *Annalen der Pharmacie* Liebig created a journal which has ever since been one of the chief repositories of the best results of the laboratories of Europe, and especially of Germany. At the time of Liebig's death, in 1873, 165 volumes of the *Annalen* had appeared, and there has been more than an equal number since that date.

The *Handwörterbuch* which Liebig, with the co-operation of his friends Poggendorff and Wöhler, produced between 1836 and 1856, and the *Handbuch der Chemie* in 1843 must also be mentioned. The famous *Letters on Chemistry* were originally published as newspaper articles in the *Augsburger Allgemeine Zeitung* with the object of bringing within the ken of the general public some of the more important consequences of the advance of knowledge in connection with the affairs of everyday life.

Again, up to 1847, Berzelius had for many years prepared annually a *Jahresbericht über die Fortschritte der physischen Wissenschaften*, but near the end of his life this laborious undertaking was no longer possible for him, and Liebig, in association with Hermann Kopp, the physical chemist, commenced the *Jahresbericht*, which, so far as chemistry and the allied sciences are concerned, continues to this day. It is no longer so important as formerly, having fallen behind in date, but for certainly forty years it was indispensable to every practising chemist who was directly or indirectly interested in the progress of the science.

Since the days of seventy or eighty years ago, when Liebig set these enterprises in motion, the number of periodical publications devoted to recording advances in chemistry has greatly increased, and a number of journals now appear at regular intervals of a month, a fortnight, or even a week, which have become necessary in consequence of the specialisation which is characteristic of our time. We have therefore journals of inorganic chemistry, physical chemistry, applied chemistry, and some limited even to one topic, such as electrolysis or radium. Liebig's *Annalen*, however, continues to hold an honoured place in every chemical library.

The creation of a school of thought, such as that of which the chemical school at Giessen was the centre, requires originality as well as learning in the teacher, intelligence in the taught, and a sympathetic relation between professors and students. These are more important than buildings and appliances. But much influence is exercised by the environment; that is, by the attitude of the public. Appreciation of learning and interest in the results of research have long been provided more freely in Germany than in England. Though we cannot now admit, without qualification, the reproach of Liebig, already quoted, it is still true to some extent that what the public in England wants is invention rather than discovery; the applications of knowledge before the knowledge itself.

Some people will doubt, perhaps, whether we are so much behind Germany, "learned, indefatigable, deep-thinking Germany," as Carlyle called her. We have an immense amount of popularisation of the results of science, but it is to be feared that much of this is too easy, shallow and misleading.

Possibly the difference between the two peoples is to be accounted for partly by the attitude of the Governments in the two countries. In England it has been the custom to leave the investigation of many important subjects, like agriculture, to the chance of private benefaction or voluntary effort. In England, again, it is only in comparatively recent times that assistance out of public funds has been given to the universities. This attitude of the Government has an immense influence in directing popular views of institutions, of things, of men. That which the masses find placed in positions of advantage by the powers set over them is naturally held in higher esteem than that which is always kept in the background or in a position of evident inferiority. In Germany the university chairs are occupied by the greatest specialists in every department, and these are men who are honoured at Court, consulted by Ministers, and trusted by manufacturers. But, after all, when we have exhausted the enumeration of all the adventitious influences at work in both countries, it seems as though there are some elements in the mental constitution of the different peoples which lead them to handle the same subject of inquiry in different ways. It has been so in the study of chemistry.

At the beginning of the nineteenth century, with the aid of the principles bequeathed by Lavoisier, the facts which had been established by Priestley and Cavendish, the discoveries of Humphry Davy, and the atomic theory of Dalton, France and England were engaged in laying the foundations of the new science. At that time Germany had no chemists. Liebig himself bears witness, in his autobiography, that in his youth "it was a very wretched time for chemistry in Germany." During the latter half of the century, however, there arose in nearly every German university a famous school of chemistry, and in practically all cases it has been a school for the cultivation of so-called "organic chemistry," in which department German chemists have achieved the most brilliant successes. Nothing can be more important than Kekulé's theory of the aromatic compounds. Nothing can be finer than the synthetical work of von Baeyer and Emil Fischer in connection with indigo, the sugars, and the proteins, or albuminoid substances, the chief basis of the animal tissues. But it cannot be maintained that they have been equally distinguished for the discovery of broad general principles. German triumphs have been more frequently the result of that patient attention to detail which seems characteristic of the German mind.

Take, by way of illustration, the problems which at the present time loom largest before the chemical world. There are first the relations among the atomic weights, discovered by Newlands, an Englishman, and worked out by Mendeléeff, a Russian; next, the arrangement of atoms in space, or stereo-chemistry, to which the clue was furnished by Le Bel, a Frenchman, and van 't Hoff, a Dutchman; next, the process of electrolysis and the constitution of salts in solution, of which by far the most important theory, the theory of free ions, was supplied by Arrhenius, a Swede. Again, there is radio-activity with all its consequences, the isolation of radium by Madame Curie, and the greater part of its wonderful history, worked out by Rutherford and Ramsay, both British chemists. Practically the whole of modern knowledge concerning the constitution of atoms is the product of investigation in British laboratories carried out by Crookes, J. J. Thomson, Rutherford, Soddy and a few others. To those great fields of inquiry Germany has, doubtless, made contributions, but she did not discover them.

In the course of his busy career Liebig was led into many a controversy, and if in the heat of discussion he was sometimes tempted into the use of expressions which seemed to indicate undue bitterness of feeling, it is not to be inferred that his nature was irascible or unjust, and that kindness and generosity were unknown to him. There seems no doubt that while he held rather tenaciously to theories of his own, his respect for truth was such that he was ready to change his view so soon as well-established facts proved him to be wrong. Among men of science it may also be said that few were more free from personal vanity, and notwithstanding the very numerous distinctions bestowed upon him by the scientific institutions and Governments of England, France and Germany—the Copley Medal of the Royal Society, the Foreign Associateship of the French Academy, the order *Pour le Mérite*—they affected in no way the simplicity of his life or his devotion to the pursuit of science. Hofmann, a former student under Liebig at Giessen, made the story of his life the subject of the Faraday Lecture given by him to the Chemical Society in 1875. In this lecture he related an incident which seems to illustrate so clearly the benevolent and kindly elements in Liebig's character that the whole of it in Hofmann's words will be acceptable to readers of this notice. He says: "Many years ago (in 1853) Liebig was making an excursion among the mountains of the Tyrol, and I and two others of his friends had the happiness of being his companions on the tour. In the course of our rambles one morning we overtook an old soldier who was travelling slowly along the road, much wasted by fatigue and obviously enfeebled by disease. As we came up to him he accosted us with a piteous tale, and humbly implored our aid. Following Liebig's example, whose purse on such occasions was ever as freely open as his heart, we made up among us a little stock of florins, which the poor man evidently regarded as a small fortune dropped by Providence into his hand; then pushing forward we soon left him behind, and in half an hour's time reached a village inn, at which we agreed to rest ourselves and dine. While thus engaged we observed our poor wayfarer also enter the inn. It pleased us to reflect that, for this once at all events, he had the means of procuring a comfortable meal; and having finished our own we resolved to take a short *siesta* before setting out again on our journey. After some half an hour's doze I awoke and found two of my companions fast asleep in their chairs, whilst Liebig, to my surprise, had disappeared. I immediately got up, and proceeding to the bar, inquired of the inn-keeper where our friend, the elderly, spare man of the party, had gone. The landlord replied that the gentleman had been inquiring, a little while ago, for a pharmacy; and that, upon learning there was none in the village, nor any nearer than in the next village over the hill, he had set out on foot in that direction. Not without some little anxiety at the temporary dispersion of our party, I at once proceeded on the road which Liebig had taken. After half an hour's walk I observed his figure on the brow of the hill, and hurried forward to meet him, impatient to learn the object of his solitary promenade. He answered me simply that he had perceived in our poor soldier symptoms of low fever, such as quinine was certain to cure, and that he had been over to the nearest pharmacy to get some of this remedy. On his arrival, he added, the apothecary chanced to be absent, but his wife had given him (Liebig) the free run of the bottles, with permission to select therefrom any article he might desire, paying, at his own price, for whatever he might take. He went on to tell me that fortunately he had discovered the quinine bottle, and made up, with a portion of its contents, a box full of powders, sufficient, he hoped, for our wanderer's perfect cure. After another half-hour's walk, the powders were delivered to the soldier, with instructions how often they were to be taken. Not a word was said of the long walk they had cost the kindly donor."

In the later years of his life Liebig suffered a good deal from bad health, and when Wöhler proposed a joint research Liebig was unable to accept the proposal. His life at Munich was sufficiently filled with occupations, partly domestic and partly in the application of scientific notions to practical purposes, such as making bread and the preparation of foods for invalids and children. In 1871 he gave an address to the Bavarian Academy of Sciences which displayed a characteristic liberality of spirit toward the French, then suffering deeply from the consequences of the war. "It will be on the neutral ground of science," he said, "that the best minds of the two nations must meet in endeavouring to reach the high goal common to both," and since the tables have been again turned by the results of the late war, it will perhaps be possible in time for the same sentiments to be revived on both sides.

During the winter of 1872 Liebig continued to give his lectures, and in January, 1873, he was still busy with experiments; but the end was then near at hand, and on April 3rd his last letter to Wöhler showed that he was suffering from sleeplessness and exhaustion. The friends never met again, for Liebig died on April 18, 1873, at Munich. Wöhler lived till 1882.

Jean Baptiste André Dumas.
From the Moniteur Scientifique Quesneville, Avril, 1880.

[*Face p.* 205]

CHAPTER XVI

DUMAS

LIEBIG'S great contemporary in France was Jean Baptiste André Dumas, whose association with science began, as in so many other cases, in connection with drugs and the apothecary's shop. Dumas was three years the senior of Liebig, having been born at the little town of Alais, in the Department of the Gard, July 14, 1800. His father came of an ancient family which at the revocation of the Edict of Nantes had separated into two branches. Of these the Protestant branch had emigrated, while the Catholic, to which he belonged, had remained in France. After several years in Paris he had settled in his native place, where he held the position of clerk to the municipality.

Alais at that time was but a small place. Nevertheless the local college provided all that was necessary for the boy's early education, including the study of the Latin language, so congenial to the classical traditions of the neighbourhood, filled as it is with splendid relics of Roman occupation. These associations could not but have a tendency to direct the mind of young Dumas to the study of the past, but there were other influences, not less potent, continually recalling to the present; for the country of his birth afforded opportunities alike of observing Nature and the application of her products to the service of man. Both in his speeches and writings in later life he frequently refers to these varied impressions derived from his early youth at Alais. Having made up his mind to enter the Navy he would have presented himself for examination but for a deficiency in some branches of mathematics. In the meantime the political events of 1814-15 obliged his family to renounce this project and select for him a career which would entail less sacrifice. He was accordingly apprenticed to an apothecary at Alais.

Finding, however, little to his taste in this avocation, and few opportunities for progress in scientific study, he determined to abandon the pharmacy, and in 1816 he started on foot for Geneva, where he had relatives. Here he found lectures on botany by de Candolle, on physics by Pictet, and on chemistry by Gaspard de la Rive. He had, moreover, the privilege of using a laboratory belonging to the pharmacy of Le Royez. The pharmaceutical students, who frequently joined in botanical excursions during the summer, now united in arranging a series of meetings for scientific study in the winter, and, seeing that Dumas had access to a laboratory, it was suggested that he should give them a course of experimental chemistry. Young as he was, he soon attracted the attention of his teachers, and he began to think of the possibility of joining an exploring expedition to some distant part of the world. With the desire to become familiar with the language and ideas of botanical science he composed a monograph on the Gentianacea, but the study of physics and chemistry, especially the papers by Davy, Berzelius, Gay-Lussac and Thénard in the *Annales de Chimie* served to provide strong attractions in another direction.

About this time he was fortunately able to render very useful service to Dr. Coindet, one of the principal physicians in Geneva, and thus found opportunities for making himself better known. Dr. Coindet asked him to examine carbonised sponge and to ascertain particularly whether iodine were present in this material. An answer in the affirmative led to the introduction of a number of preparations of iodine and iodides into medical practice. The new remedies being mentioned in a German journal published at Zürich in this connection, the name of Dumas first appeared in scientific literature.

About this time Dr. J. L. Prévost, after an absence of several years, returned to Geneva. He had long been absent in Edinburgh and Dublin devoting himself to comprehensive studies in several departments of medicine. He induced the young chemist to join him in physiological researches relating especially to the composition of blood. Their first memoir on the subject appeared in the *Bibliothèque Universelle de Génève*, and in the title Dumas still figured as *Elève en Pharmacie.* Other physiological problems were also studied, and this fact serves to explain the readiness with which Dumas in the later years of his career

delight; a new horizon began to dawn upon me. Save the time devoted to some visits, I was allowed to remain the whole day with Humboldt, who darted from point to point over the vast range of his recollections whilst I endeavoured to keep pace with the uninterrupted flow of his narrative. Sometimes the mountain scenery would remind him of the Cordilleras, though it must be confessed he did not think much even of Mont Blanc. Sometimes he turned to science and then astronomy and physics, chemistry and the natural history branches would in rapid succession come in for their share in the dialogue, which, spoken in a low, somewhat monotonous tone, would have scarcely appeared impressive had it not been for some waggish pleasantry which now and then escaped as it were involuntarily."

From this remarkable man Dumas learned the character of the scientific world in Paris, and the opportunities which would be open to him there of gaining the advice and assistance which would enable him to carry out the researches in which he desired to engage. The consequence was that Dumas soon afterwards found his way to Paris. Arrived in the capital, he had the good fortune very soon to make the acquaintance of three young men of about his own age, all ardent students of science. These were Victor Audouin, the zoologist, Adolphe Brongniart, already well known as a botanist, and Henri Milne-Edwards, then working for his medical degree. The friendship of these three men was always regarded by Dumas as a beneficent influence during the earlier years of his career. Moreover, this intimacy led to acquaintance with Herminie Brongniart, sister of his friend Adolphe, and eldest daughter of the distinguished geologist. On February 18, 1826, the marriage proposed was concluded and Herminie Brongniart became Madame Dumas. A devoted mother of their son and daughter, a counsellor and helpmate, she remained to the end of his life a sharer in all its vicissitudes.

A kindly feeling of sympathy with young workers in the field of science was a notable feature of the society of men of science gathered in Paris at that period, and young Dumas found himself admitted in a short time to the intellectual life of the capital. Laplace and Arago, the famous astronomers, Berthollet, Vauquelin, Gay-Lussac, Thénard, leaders among the chemists, Alexandre Brongniart, Cuvier and Geoffroi Saint-Hilaire, distinguished among naturalists, Ampère and Poisson, eminent as physicists, were among this company in which Dumas soon found a place. The position of *Répêtiteur de Chimie* to Thénard's course of lectures in the École Polytechnique became vacant and, on the proposal of Arago, Dumas was appointed. Soon afterwards the professorship of chemistry at the Athenæum, an establishment for providing evening lectures somewhat resembling the Royal Institution in London, fell vacant and Dumas was appointed. These two engagements provided him with full occupation, and for a time his research work was interrupted, while the study of physiological questions receded into the background and chemical problems occupied his attention. He now began to collect materials for his grand *Traité de Chimie appliquée aux Arts*, the first volume of which appeared in 1828.

One of his earliest and most important works published in the *Annales de Chimie et de Physique* for 1826 related to the Atomic Theory, in which for the first time the doctrines of Avogadro and their relation to the Atomic Theory of Dalton received notice. "I am engaged," he says, "in a series of experiments intended to fix the atomic weights of a considerable number of bodies, elementary or compound, by determining their density in a state of gas or vapour." Dumas, in fact, realised the distinction between molecules and atoms, a conception which lies at the base of all modern ideas of chemical constitution. The method of estimating vapour densities associated with the name of Dumas has been of great value in a variety of ways, and is occasionally practised down to the present day. One result of his enquiry into the atomic weights of the elements by this method was the settlement of the composition of silica, and consequently of the constitution of all the numerous natural silicates. This enquiry and the conclusions which he was led to form brought him into conflict with the great Swedish chemist, Berzelius, then at the height of his fame and position of authority in connection with the science of chemistry. Dumas' view, however, prevailed, and the idea that the molecule of silica consists of an atom of silicon combined with two atoms of oxygen has never since been questioned.

At the time referred to "organic" chemistry was in its infancy. A considerable number of definite chemical compounds, such as sugar, alcohol, ether, a few vegetable acids and bases were known, but as already mentioned the various processes introduced for the purpose of determining their composition were far from satisfactory, and science was still waiting for the method of combustion perfected by Liebig. Moreover, the relation of such compounds one to another was generally unknown and the existence of homologous series, and the phenomena of isomerism were not yet recognised. Questions of this kind soon began to occupy the attention of Dumas, and after 1830 he produced a large number of papers devoted to the composition, nature and chemical relations of compounds of carbon. The most momentous of his discoveries in this field was the establishment of the theory of *substitution,* according to which it is possible for chlorine, and the other halogens, to take the place of hydrogen atom for atom in carbon compounds. This discovery originated in a remarkable manner. The story goes that at a soirée in the palace of the Tuileries the guests were much annoyed by irritating fumes proceeding from the wax candles used, which burned with a smoky flame. Alexandre Brongniart, in his capacity as chemist to the King's (Charles X.) household, was consulted about this unpleasant incident. He entrusted to his son-in-law the task of examining the candles. Dumas was all the more inclined to engage in the enquiry, having been already asked by a merchant to suggest a method of bleaching certain kinds of wax which, resisting the ordinary methods, remained unsaleable. The irritating vapours turned out to be hydrochloric acid arising from the use of candles of which the wax had been bleached by the action of chlorine. The amount of chlorine absorbed was considerable, and experiments proved that this element was capable of entering in a similar manner into the substance of many other organic compounds. The idea that an atom of the negative element, chlorine, could replace an atom of the positive element, hydrogen, was of course repugnant to the ideas of those, and they were the great majority, who had accepted Berzelius' electro-chemical theory of combination ; and though familiar enough to every chemist in later times, this is still a phenomenon which deserves to be called remarkable. Dumas' views were received with scorn and, on the part of the German chemists, with ridicule, which was carried in some cases beyond the bounds of fair controversy. Wöhler, the friend of Liebig, and professor at Göttingen, wrote a letter which purported to come from a certain S. C. H. Windler in which one passage was as follows "The last great discovery from Paris shows that it has been found possible to replace in acetate of manganese first the atom of hydrogen by chlorine, then the oxygen, then the manganese, and at last even the carbon so that a body was formed which contained only chlorine, but retained still the properties of the original substance." Such a travesty of Dumas' views was not justifiable, and Liebig, who published the letter, found reason before long to acknowledge that he was mistaken, for the accumulated facts soon formed an irresistible phalanx against which neither authority nor theory, neither prejudice nor sarcasm could provide effectual opposition. Further study and application of the doctrine of substitution was gradually absorbed into the later doctrine of types with which the names of Williamson and Gerhardt are especially associated and which will be referred to on a later page.

The activity of Dumas knew no bounds, and it would be impossible to enumerate the subjects with which he occupied himself; it would, moreover, be confusing to the lay reader.

The number of elements employed in the composition of organic substances is very small, and for the majority is limited to carbon, hydrogen, oxygen and nitrogen. From the days of Lavoisier repeated attempts had been made to devise satisfactory processes for determining exactly the proportions of each of these elements in a given material, but it was reserved for a comparatively late period that a method was contrived in the famous laboratory at Giessen for the estimation of carbon and hydrogen. About the same time in Dumas' laboratory in Paris a practical process was introduced for estimating the nitrogen in similar compounds. There are no two chemists of the past who contributed more to the knowledge of organic compounds than Liebig and Dumas, and it is satisfactory to remember that in the language of the laboratory the name of Liebig is always associated with the process of combustion by which the carbon and hydrogen of such a compound is determined, while that of Dumas is attached to the chief method of estimating the total nitrogen.

While largely concerned with the study of organic—that is carbon—compounds, Dumas at one time engaged in the determination of some of the most fundamental quantitative relations among the combining proportions of the elements. Together with Stas, who afterwards became the great authority on this subject, Dumas carried out a series of very accurate experiments on the composition of carbon dioxide or carbonic acid, as it is commonly called. By burning diamond, the purest known form of carbon, in a stream of oxygen gas, it was shown that 12 parts of carbon unite very exactly with 32 parts of oxygen, forming 44 parts of carbonic acid, a result practically identical with that obtained by the combustion of graphite, and confirmed by the numerous later experiments made by other chemists. The determination of the exact composition of water was the subject of another series of experiments by Dumas, carried out with great skill and care. The result was of fundamental importance, for although it had been proved by Cavendish forty years earlier that water is formed by the combination of two volumes of hydrogen with one volume of oxygen, the relative densities of the two gases were not known with sufficient accuracy, nor indeed were the *exact* proportions by volume in which they unite, to provide trustworthy data for calculating the composition of water by weight. The work of Dumas was embodied in a paper which must always be regarded as classic, and should be read in the original by every serious student of chemistry.

Associated with Boussingault, Dumas also investigated the composition of atmospheric air, and discussed in an interesting manner the relation of the various agencies which in the operations of nature affect its composition.

One consequence of the establishment of the atomic theory introduced into chemistry by Dalton was the necessity of attempting the accurate determination of the relative masses of the atoms of the respective elements. We have seen how this task was begun by Dalton himself, with, however, imperfect success. We have also seen how the major part of the life of Berzelius was given to the work and how deeply science is indebted to the labours of the great Swede for the degree of accuracy he succeeded in introducing into the methods then available, in face of immense practical difficulties. A series of numbers had thus been accumulated which had been the subject of much speculation as to their significance. The most notable hypothesis was that of Prout, who in 1816 had put forward the idea that the weights of the atoms of the elements are integral multiples of that of hydrogen, the element which enters into combination in the smallest proportion of all and which in the gaseous state is the lightest substance known. The atomic weight of hydrogen was therefore taken as the unit.

To Dumas we owe not only estimations of the relative atomic

weights of hydrogen, oxygen and carbon already referred to, and those of silver and a number of other elements carried out in later years, but he was one of the first to recognise some of the numerical relations which are discoverable among the atomic weights of the elements.

Chemistry at the end of the eighteenth century consisted of an immense collection of facts in the midst of which here and there could be recognised relationships which for want of more complete knowledge were often puzzling. The discovery of new elements and the careful study of those already known gradually served to elucidate these relationships and to lead on to a more complete classification of both elements and compounds. The metals had long been distinguished from the non-metals, unfortunately called "metalloids" about 1811 by Berzelius. These latter, very various in physical as well as in chemical properties, and usually presenting no resemblance to metals, were divided by Dumas into five families, as follows:

1. Hydrogen.
2. Fluorine, chlorine, bromine, iodine.
3. Selenium, sulphur, oxygen.
4. Phosphorus, arsenic, nitrogen.
5. Boron, silicon, carbon.

The only change brought about by time occurs in the case of boron, which is no longer placed in the same family with silicon and carbon. The metals in like manner admit of being arranged in families which are usually composed of three members obviously presenting closely similar characters. The following afford sufficient examples:

Calcium	Magnesium	Lithium
Strontium	Zinc	Sodium
Barium	Cadmium	Potassium

Many attempts were made to trace the relations among the atomic weights in the several recognised families, but no general discussion of the subject appeared till 1858, when Dumas drew attention to the analogy between the series of closely related elements and the various known homologous series of compound radicles among carbon compounds. Since that day, however, many new elements have been discovered, the atomic weights have in many cases been corrected, and the whole have been brought within one comprehensive scheme which will be described later in connection with the name of Mendeléeff.

Enough has been narrated to show the extraordinary activity of Dumas in the pursuit of new knowledge, the diversity of the subjects which successively received attention from him, and it is therefore easy to explain the position of authority to which he speedily attained. The duties of the two offices to which he was appointed at the École Polytechnique and the Athenæum were heavy enough for any ordinary man. But Dumas was endowed with more than ordinary capacity and appetite for work, and joining in the project for the establishment of the École Centrale des Arts et Manufactures, he became lecturer on chemistry in the new institution.

In 1832 Gay-Lussac retired from the chair at the Sorbonne, and Dumas succeeded to this position, which he retained till 1868. From the post of Répétiteur at the École Polytechnique to which he was originally appointed he was advanced to that of professor on the retirement of Thénard in 1835. Finally he became professor in 1839 at the École de Médecine, so that in turn he took part in the work of all the great teaching establishments in Paris, including the Collège de France, where, however, he was engaged only temporarily.

There appears to be no doubt that owing to the care he bestowed in early days at the Polytechnic and elsewhere on the preparation of his lectures and the experimental illustrations, Dumas attained a distinction in style and an impressiveness which was recognised by all the diversity of audiences, whether students or members of the Academy, which in the course of his long career he was called on to address. Of those, now only a few survivors, who had the privilege, in 1869, of listening to his delivery of the first Faraday Lecture, in the Royal Institution in London, none would be likely ever to forget the occasion. The subject consisted not merely of an eulogium of Faraday; the lecturer proceeded to survey the entire field of physical science, concluding with a summary of his own views on the nature of "organic" substances and their relation to living matter. Notwithstanding the assertions which have been made by some chemists and physiologists since that day, it is still true that "the chemist has never manufactured anything which approximately or remotely was susceptible of even the appearance of life. Everything he has made in his laboratory belongs to dead matter; as soon as he approaches life and organisation he is powerless."

Such a discourse can be only imperfectly rendered in any translation; and no reproduction is possible of the noble eloquence which owed so much of its charm to the personality of the speaker.

Dumas was the first chemist in France to adopt the system of practical instruction in the laboratory introduced by Liebig at Giessen. From about 1832 he began this kind of teaching at the École Polytechnique, but after a few years he carried on the work at his own expense in his laboratory at the Rue Cuvier. This, however, came to an end as one of the consequences of the revolution in 1848, and the considerable reduction of his income. Another consequence was the necessity in which he found himself of yielding to the public demand for his services in connection with public affairs, and as a result he was elected member of the National Legislative Assembly. Subsequently he was appointed Minister of Agriculture and Commerce, a member of the Senate, President of the Municipal Council of Paris and Master of the Mint. It was inevitable that his scientific work should suffer, and from this time forward his memoirs on scientific subjects appeared at longer intervals and consisted for the most part of reports on various practical subjects.

With the fall of the second empire Dumas' political and administrative career came to an end, but though now at the age of seventy, when most men look to the association of *otium cum dignitate*, Dumas still eagerly joined in various movements for the promotion of science. He became in 1872-3 Chairman of the Commission for making preparations for observation of the transit of Venus in 1874, and he was chiefly instrumental in the establishment of the *Association Française pour l'Avancement des Sciences*, in promoting which he pointed out in an eloquent speech at Clermont the advantages to be derived from following the example of the British Association already long established.

In the autumn of 1883 Dumas' health, which had been unimpaired up to that time, began to fail, and by the advice of his physicians he passed the winter in the South of France. He died at Cannes, April 11, 1884.

Group IX

THEORIES OF CHEMICAL ACTION AND CONSTITUTION OF MOLECULES

FRANKLAND (1825-1899)
WILLIAMSON (1824-1904)

CHAPTER XVII

FRANKLAND

As already stated, the establishment of the Atomic Theory in its application to chemical phenomena resulted less from the work of the great philosopher of Manchester, John Dalton, than from the exact investigations of atomic ratios or atomic weights by the Swedish chemist Berzelius. It was long, however, before it was accepted in an unqualified way by chemists in general, and for many decades was only regarded as "at best but a graceful, ingenious, and in its place useful hypothesis."[1] In the meantime the employment of chemical formulæ, with the symbols introduced by Berzelius, was universal, and efforts were constantly made to arrange these symbols in some way to recall the chief properties and reactions of compounds. This, however, could never have become possible but for the discovery of the principle on which the idea of atomic valency is based. This principle with its remarkable consequences in every department of theoretical chemistry was first announced by Frankland in 1852.

Edward Frankland was born January 18, 1825, at Churchtown, near Garstang, in Lancashire. He has left a volume of reminiscences which render it comparatively easy to trace his development from childhood and his progress through a long career of usefulness and scientific distinction.

Frankland's life affords another example of the fact that to a man of genius the amount and character of any systematic education he may have received has but a comparatively small influence on his success in the career he has marked out for himself. His experience at the village schools he first frequented was similar

[1] *Fownes' Manual*, 1847.

Believe me
Yours very truly
E. Frankland

to that of most other children in those days. He must have been somewhat precocious, for he begins by saying: "I cannot recollect when I could not read, but have been told that I knew the alphabet at the age of two years. When three years old I was sent to a Dame's school at Manchester, where my mother was staying for a short time with her sister." He records his indebtedness to the master, Mr. James Willasey, under whose tuition he remained, from the age of seven, for five years. "Mr. Willasey was a real educator, not a mere schoolmaster." It was while at this school that young Frankland learned not only to use a good accent in speaking French but he became "interested in chemistry, electricity and magnetism, and read many books on these subjects, particularly Priestley's electricity." He also made a Voltaic pile, and invited by his master, Mr. Willasey, he exhibited his apparatus and experiments before the whole school. His friendship with Mr. Willasey was kept up to the day of Willasey's death, and after the school had to be given up he assisted in providing for the old man, who had fallen into poverty. The last school Frankland attended was the Lancaster Grammar School, a school of the old classical type concerning which he makes in his recollections several notable remarks. The use of the cane was customary, and with regard to this corporal punishment he expresses the opinion that very few of the boys would have learned so much as they did without this stimulus. He hated Latin, but aided, apparently, more by the fear of the cane than by intellectual interest, he made in three years surprising progress, and not only read a good deal of Cæsar, Ovid, Virgil and parts of Horace and the plays of Terence, but achieved distinction in the manufacture of Latin verse.

"I can imagine," he says, "that intelligent commentaries upon what was read might have made the acquisition of Latin interesting to me; but as it was those three years spent upon it were the most dreary and tiresome of my life; whilst the knowledge acquired has been of extremely little use to me; for although I continued the study of the language for some time after leaving school I could never read a Latin author with pleasure, or indeed without the painful and laborious use of a dictionary."

Of the practices connected with the Lancaster school in Frankland's day (1837-40) presumably most have disappeared; the right of the sixth form to solicit money from the bridegroom at every wedding in the adjoining parish church, the raffling for prizes in the school without examination, and irrespective of merit, and the witnessing executions at the Castle.

In the choice of a calling it had been suggested by Mr. Willasey that the boy should enter the profession of medicine, but although his parents were willing, the cost appeared to them prohibitive, and it was therefore decided that the only entrance available "was the back door of a druggist's shop. This was the only mistake of any consequence," he says, "that has occurred in my career, but it was for me a very serious one. It condemned me to six years' continuous hard labour, from which I derived no advantage whatever, except the facility of tying up parcels neatly."

The selection of the master to whom he was to be apprenticed was determined, according to Frankland, by his reputation for "piety." But this apparently was not inconsistent with neglect of his apprentices and failure to teach them their business. In his recollections Frankland complains bitterly of this conduct, but he found friends in Mr. Christopher Johnson and his son, Dr. James Johnson, who lent him books and apparatus and went so far as to provide a small laboratory in which he and his friends among the apprentices in the town gathered on two or three evenings in the week for the repetition of practical exercises in chemistry and physics. Frankland's circumstances as an apprentice appear to have been less fortunate than were those of other chemists who began life in the same way in the druggist's shop, and of whose career a record has been preserved. The business in Lancaster appears to have been of that mixed character which is still common in many country places, and the supply of paints and groceries formed a large proportion of the daily transactions. The duties of the apprentice during the first two years were at 5.45 in summer and 6.45 in winter to fetch the key from his master's house, to open the shop door and take down the shutters, sweep out the shop, dust the bottles, and in winter light a fire before the master arrived at 8 o'clock. Frankland's first piece of work was to wheel a cask of treacle through the streets to the shop of a customer in the suburbs. The day seems to have been spent chiefly in pounding drugs in the cellar. The occupations must be imagined, but after two years there was considerable amelioration of the hardship, but with no more intellectual advantage. The junior apprentice succeeding

Frankland was George Maule, afterwards of the firm of Simpson, Maule and Nicholson, the celebrated aniline colour manufacturers. He lived to make a large fortune out of the colours. The other older apprentice was Robert Galloway, who in later years became Professor of Practical Chemistry in the Government School of Science in Dublin.

It says much for Frankland's ardour in the pursuit of knowledge that notwithstanding the long hours and hard work which burdened the years of his apprenticeship he frequently got up at 4 o'clock in the summer months to join a companion in botanical excursions into the fields. Thus in addition to the experience gained in experimental work under the guidance of Dr. James Johnson, as already mentioned, at the end of the six years as apprentice he was not entirely ignorant of the rudiments of several branches of science.

On the termination of his apprenticeship, of which three months were remitted, Frankland went to London in October, 1845, armed with letters from his friends the Johnsons. By their advice he entered the laboratory of Dr. Lyon Playfair, who had just been appointed chemist to the Government Department of Woods and Forests. Playfair, being occupied with official duties elsewhere, was seldom present, but Frankland records his indebtedness to Mr. Ransom, the chief assistant, a particularly amiable instructor by whom he was inducted into the mysteries of chemical analysis. "It was now," he says, "that my chemical training really commenced." At this time he lived with an uncle and aunt in Lambeth, and though not definitely so stated, this transference direct from the druggist's shop to a scientific laboratory must have been effected with the consent of his parents. His progress in analytical work was so rapid that at the end of six months Dr. Playfair offered him the post of lecture assistant at the Civil Engineering College at Putney, where he held the post of lecturer on chemistry. After he had occupied this position about six months he received a tempting offer of a place at the Royal Agricultural College at Cirencester.

In the meantime, having made the acquaintance of Hermann Kolbe (afterwards a famous professor in Germany), who had come over as assistant in Playfair's laboratory, Frankland determined, much to the surprise of his friends, to leave the affair at Cirencester, and yielding to Kolbe's persuasion, he decided to go with him to Marburg to work under Professor Bunsen,

In the long vacation of 1846, on a visit to his parents in Lancaster, Frankland made the acquaintance of Mr. George Edmondson, the proprietor of Tulketh Hall School near Preston. A very short negotiation led to his engaging to become, on his return from Germany, science master at the new school he was about to establish at Queenwood, Hampshire. In May, 1847, Frankland and Kolbe set out together for Marburg, crossing from Dover to Ostend and *via* Cologne up the Rhine, and finally by diligence. The morning after arrival they began work in the laboratory, where they were received by Professor Bunsen with extreme kindness. Here Frankland learned the processes of gas analysis from the inventor of them himself, who also taught him how to make the graduated glass tubes (eudiometers) and other apparatus required. He also joined Kolbe in a continuation of the enquiry they had commenced in London, *viz.*, the synthetical production of the acids of the acetic series, and the results of which they communicated to the Chemical Society of London.

The three months spent in Marburg were very happy, and there within a few days of his arrival young Frankland met the lady, Fraülein Fick, who two years later became his wife. The time spent at Marburg on this occasion was but short, as he was urgently requested to come immediately to Queenwood. On his arrival he found the work very hard, as he was required to teach not only chemistry, at the same time organising the laboratory, but to give lectures on geology and botany. He derived, however, two great advantages from association with the school: the one was practice in lecturing, the other was making acquaintance with John Tyndall, who had arrived a few weeks earlier.

Tyndall, who afterwards became the famous professor of physics at the Royal Institution, at that time had little knowledge of experimental science, while Frankland required instruction in mathematics. The two young men struck up a warm friendship and agreed to be mutually helpful. So by rising at four o'clock in the morning they made time for study, and while Frankland was at work at algebra and Euclid under Tyndall's direction the latter was occupied with a systematic course of qualitative chemical analysis. During this time the isolation of the alcohol radicals methyl, ethyl, etc., was the subject which chiefly occupied Frankland's mind, and according to an entry in his diary the first experiment in this direction was tried on April 10, 1848. After a number of fruitless experiments the action of zinc on the iodide of the alcohol radical was tried, with results which seemed promising, and the process further examined when Frankland returned to Marburg led to success.

On June 15, 1848, Frankland and Tyndall started for Paris by way of Southampton and Havre. They stayed a few days in Rouen, and on the 19th they left for Paris. Tyndall left Paris on June 23rd, but Frankland, remaining, had the thrilling experience of witnessing many of the scenes of street fighting during the brief revolution. He got opportunities also of hearing Dumas lecture, also Frémy and several professors at the Sorbonne. Both he and Tyndall derived much advantage from listening to lectures by the most accomplished and experienced lecturers, and in seeing the experiments used by way of demonstration. Thus they both became acquainted for the first time with the use of electric light for lecture illustration. It was, of course, in those days produced by the ignition of carbon points connected with the terminals of a Bunsen primary battery.

In October, 1848, Frankland returned to Marburg in company with Tyndall. He records as follows the sum of the advantages which he felt he had received from his connection with Queenwood: "It gave me a greatly increased knowledge of mathematics, a fairly good knowledge of geology and botany (acquired by lecturing on these subjects; there is nothing like lecturing on a subject for getting to know it thoroughly), the power of expressing myself fluently in lectures, shorthand (which was afterwards of some service in taking notes of lectures in Germany), and lastly practice in research. . . . Had my appointment in Queenwood been more congenial, so that I could have continued to hold it for a longer period, my research work would, I doubt not, have proceeded on the same lines as in Marburg, inasmuch as I had already made the fundamental reaction of zinc upon iodide of ethyl; but the progress of the work would of course have been very much slower."

The results of all his research work continued and completed in the Marburg laboratory furnished the material for the Dissertation presented in the summer of 1849 to the Philosophical Faculty of Marburg with a view to the Ph.D. degree. The examination as usual was entirely *vivâ voce* and the disputation was dispensed with, on the ground that as a foreigner he had not sufficient command of the language. It had previously been in Latin. Within a month of taking his degree Frankland became engaged to his first wife, to whom he attributes much of his success in the appointments he subsequently held during twenty-three years of happy married life.

Fraülein Sophie Fick was the second daughter of Oberbaurath Doctor Fick, of Cassel, but she was at this time visiting her brother Ludwig, professor of anatomy in the University of Marburg. She was probably the only lady in the place who spoke English, and at this time Frankland's knowledge of German was but slight.

He now decided to proceed to Giessen, and after an interview with Liebig he entered the famous laboratory in which most of his contemporaries among English chemists, Playfair, Gregory, Fownes, Williamson and Kane, beside notable Germans, among them Hofmann, Will and Strecker, had received the whole or part of their chemical instruction. As already explained, this was classic ground, being the first laboratory in the world opened (1824) for systematic instruction in chemistry and is further celebrated on account of the number and importance of the researches carried out there.

Frankland continued his work on the action of zinc on the alcoholic iodides, this time selecting amylic iodide obtained from the fousel oil of fermentation. The results of work at Giessen were published in the Journal of the Chemical Society under the title "The Isolation of Amyl." It may, in passing, be remarked that the substances at that time supposed to be the radicals methyl, ethyl, and amyl separated by Frankland from the iodides consisted really of compounds having the composition of these radicals but double the molecular weight, expressed by the formulæ $(CH_3)_2$, $(C_2H_5)_2$, etc. They were in some respects therefore, of less theoretical interest than the *organo-metallic* compounds $(CH_3)_2Zn$, $(C_2H_5)_2Zn$ and others obtained in the course of the same reaction. The remarkable compound called cacodyle had been discovered by Bunsen in 1839, and the zinc compounds produced by Frankland were manifestly of similar nature, as also were the tin compounds studied by him after his return from Germany. On consideration of all the facts Frankland was led to recognise the principle that the combining capacity of each element is limited, and that the combining power in any given case is always satisfied by the same number of atoms. This is the basis of the law of valency which is the essential and fundamental idea on which all theories of chemical constitution are founded. This constitutes Frankland's greatest contribution to chemical science, for although he pursued research successfully in many other directions, no other discovery of his has affected the progress of scientific chemistry, in the broadest sense, to any similar extent.

After a semester in Liebig's laboratory it was proposed to Frankland that it would be well for him to work for a time in the laboratory of H. Rose, the great analytical authority in Berlin, and by Liebig's intervention he obtained an offer of this privilege. But before his intention could be carried out he received an offer of the professorship of chemistry hitherto held by Playfair at the Putney College. Influenced by his anxiety to fulfil his engagement with Sophie Fick, this offer was accepted, but his stay at Putney was short, for in 1850 the foundation of the Owens College at Manchester was announced, and in 1851 Frankland was engaged as the first professor of chemistry. He presented an array of testimonials such as few young men of eight and twenty then, or in more recent times, would be in a position to display, but the emoluments of the new post were modest enough. The salary attached to the chair was £150 per annum, with two-thirds of the fees of the students, apparently without any guarantee as to a minimum, but it was sufficient to determine the immediate marriage of the two young people. In consequence of legal difficulties Sophie Fick was brought over from Cassel by her brother Heinrich, and the marriage took place at St. Martins-in-the-Fields on February 27, 1851.

The history of the Owens College, long since become the nucleus of the University of Manchester, is now a familiar story, but in so far as it concerned the career of Frankland the early days were full of difficulty. By October, 1851, the lecture and laboratory courses were arranged and a little time was found for research, but this was frequently interrupted by numerous lecture courses of special character, both in Manchester and London.

But the Manchester people in the early fifties were not ready for teaching of university type, and in many of the departments of the College but few students remained. An opportunity of release happily occurred in the vacancy at St. Bartholomew's Hospital, to which Frankland was appointed in succession to Dr. John Stenhouse, in 1857. To this post he also added in 1859 the lectureship at Addiscombe Military College, and he was appointed in 1863 successor to Faraday at the Royal Institution. The number of lectures to be given in the course of a week was, therefore, considerable, and as Frankland began at this time to be employed by manufacturers and in the law courts as a scientific expert, it is somewhat surprising that under the stress of so many duties and engagements his health did not break down. After a time he gave up the lecturing at Bartholomew's and at Addiscombe, and confining himself to work at the Royal Institution, he was able to accomplish a good deal of original research, the results of which appear in papers communicated to the Royal and the Chemical Societies.

He had always been fond of travel, and his excursions to Germany in search of instruction, and to Paris, where he found adventure enough, in a city where a revolution was in progress, have already been mentioned. Immediately after his marriage he visited Tenby, accompanied by his wife, for the purpose of a technical investigation of certain anthracite coal. In the following summer they went to Cassel, but after the arrival of children a cottage at Windermere was secured in which during the succeeding three or four years accommodation was found there not only for the young family but for his father and mother.

Frankland was very fond of yachting and had a small yacht on the lake ; and later on, at Cowes, I.W., he kept a cutter just big enough for two people to spend the night in. Letters to his wife in later years show how much pleasure he derived from this sport, in which he enjoyed the company of some of his scientific friends and occasionally of his wife and children.

In 1859 he undertook, in company with Professor Tyndall, a journey with a scientific object, and of a more serious character. Tyndall had obtained from the Royal Society a grant of money for the purpose of establishing on Mont Blanc a series of thermometric stations and Frankland gladly accepted his invitation to join in the expedition. They started from Chamounix with the intention of spending at least one night on the summit and accordingly were provided with a tent in addition to the posts for the thermometers and other apparatus. The whole party on starting numbered thirty-one persons. Of these a number of the porters were dismissed as soon as the summit was reached, but those remaining suffered somewhat severely from mountain sickness. Tea was the only form of refreshment desired by any member of the party, and for solid food there was no desire. The scenes on the way and at the summit have been so often described that it is unnecessary to attempt any account of the wonders of the sunset and sunrise as depicted by the travellers on this occasion. While Tyndall was busy erecting the posts for his thermometers Frankland collected samples of air for analysis and occupied himself with experiments on the burning of stearine candles, the rate of which had been tested in the valley below. It was found as a fact, contrary to assertions which had been made by an American physicist on theoretical grounds, that the candles in burning one hour on the top of the mountain were reduced in weight to almost exactly the same extent as when burnt for the same time in the valley. An unexpected phenomenon was, however, revealed, for it was noticed that the amount of light given out by the candles was greatly reduced, the blue non-luminous part of each flame extending much higher than usual. Experiments repeated afterwards in England confirmed the conclusion that the illuminating power of hydrocarbon flames is directly proportional to the pressure of the atmosphere in which they are burning. This observation led to a long series of researches culminating in a new theory of the luminosity of flames in general.

In 1863 Frankland for the first time visited Norway, the country he loved so much, and where thirty-six years later he died. The letters home contain much interesting detail concerning the scenery and physical phenomena connected with the glaciers and other features of the country, also remarks on the people and the charm of travel in a country then comparatively unsophisticated. Later visitors to Norway have not found the same comfort, economy and enjoyment as characteristic of the people and country as in 1863. Frankland roamed up and down the coast, visiting the Lofoden Islands and the North Cape in the course of these excursions. He also enjoyed much salmon- and trout-fishing.

In 1865 he succeeded Dr. Hofmann as professor in the Royal School of Mines at South Kensington and the Royal College of Chemistry, then in Oxford Street. In 1868 he became a member of the Royal Commission appointed to enquire into the pollution of the rivers and water-supply of Great Britain. The work in which he thus became involved occupied much of his time during the six years when the water laboratory remained in operation.

From the time of his return from Manchester in 1857 till 1870 Frankland lived in Park Road, Haverstock Hill. But in

Q

that year he removed to 14, Lancaster Gate, Hyde Park. About this time his wife's health, which had been failing for some years, became so much worse that she was advised to go to Switzerland. She died at Davos on January 7, 1874, leaving two sons and two daughters. Her second son became the well-known professor in the University College of Dundee, and afterwards for twenty-three years in the University of Birmingham.

Frankland married again in 1875 Miss Ellen Frances Grenside, and in 1880 he bought a small estate, The Yews, at Reigate, where he had a large garden, and where he built himself an observatory.

That Frankland was a good manipulator and glass-blower may be inferred from the character of his early researches, which involved the frequent employment of sealed glass tubes holding gases under pressure, and much experiment in the analysis of gases. The establishment of his observatory on Haverstock Hill led to his attempting with ultimate complete success the grinding, polishing and silvering of specula for the telescope. At Reigate he set up an electrical installation for lighting the house, many of the fittings of which he made and fixed with his own hands. He was always interested in gardening, and even at Haverstock Hill he kept a greenhouse well stocked with flowers and grew abundance of fruit.

A section of Frankland's autobiography is headed "Religion," and he there traces with unusual detail and in an interesting manner the gradual change in his views from those with which his youth was associated. He says: "I was duly baptised into the Church of England and at the age of two years was taken regularly by my mother every Sunday to Churchtown Church, where we sat in the big square pew of my great-uncle, John Dunderdale. . . . When my mother married William Helm we went to the Congregational Chapel, in Lancaster, as my stepfather was a Congregationalist." His experiences during his youth were similar to those of many a thoughtful young man who finds the orthodox views intolerable. Even at the age of twenty-two, when he went to Marburg, he was distinguished among the students by going to church. But after returning to England he became gradually convinced of the untenable character of so much that he had been taught that he became confirmed in his agnostic attitude.

In 1863, when he was much older, he joined a small number of earnest scientific workers who met once a month and dined together. This company, under the name of the "X Club," consisted of G. Busk, T. A. Hirst, J. D. Hooker, T. Huxley, J. Lubbock, Herbert Spencer, W. Spottiswoode, John Tyndall, E. Frankland. It was never added to and was only gradually extinguished by death. "All these colleagues of mine," wrote Frankland, "occupied some of the highest positions in the scientific world, and were of one mind on theological subjects."

In January, 1899, Lady Frankland died, and the shock of her death affected her husband deeply. However, he went in the summer to his beloved Norway and there occupied himself in dictating to his secretary his recollections of travelling. He was, however, seized with illness which ended fatally on August 9th. He was buried in Reigate Churchyard.

Frankland was created K.C.B. in 1897 on the occasion of the Queen's Diamond Jubilee. It is superfluous to enumerate the many British and foreign distinctions conferred on him. It is sufficient to say that beside a Royal medal in 1857, he received in 1894 the Copley Medal, which is the highest honour at the disposal of the Royal Society.

CHAPTER XVIII

WILLIAMSON

ALEXANDER WILLIAMSON lived till the beginning of the twentieth century, but the work which he did in chemistry was limited to a period of a few years in the middle of the nineteenth. He retired from the chair which he had held for nearly forty years at University College, London, in 1888, and no paper, on researches conducted by himself, had been published for many years. Hence it is not surprising that the name of Williamson is little known to students of chemistry at the present day. Nevertheless the influence of his ideas in the development of modern theoretical chemistry is incontestable. He was also a sturdy and consistent supporter of the Atomic Theory at a time when, in the middle of the nineteenth century, there were many indications that the idea of atoms as physical entities had not been completely assimilated and assented to by chemists generally.

In what follows advantage has been taken of two biographical notices which have been written by two friends, now both deceased; the one in the *Proceedings of the Royal Society*, by Professor Edward Divers, F.R.S., the other in the *Transactions of the Chemical Society*, by George Carey Foster, F.R.S., formerly professor in and Principal of University College, London; both were intimately acquainted with Williamson.

Alexander William Williamson was born at Wandsworth on May 1, 1824. His father, Alexander Williamson, came from Elgin as a boy, and in 1820 married the daughter of William McAndrew, a Scotsman settled in London as a merchant. Mr. and Mrs. Williamson had three children, Antonia Helen, born in 1822, Alexander William, and a second son, James, who died in childhood. Antonia married a Mr. Clark and died a widow several years before her brother. Like her eminent brother,

[Face p. 228]

Antonia displayed considerable force of character and intellectual independence. Williamson throughout his boyhood was in delicate health, and though by the time he was about sixteen his condition improved, one eye remained permanently useless and the left arm was almost powerless. It was in spite of these serious disadvantages that all his experimental work was accomplished.

Between 1825 and 1831 Mrs. Williamson and her children spent much time at Brighton, where Mr. Williamson, then a clerk in the East India House, paid them frequent visits, notwithstanding the somewhat long coach journey each way. Subsequently Mr. Williamson bought a house and garden in Wright's Lane, Kensington, long since built over. His superior officer in the East India House was James Mill, the father of John Stuart Mill, and as the families lived near together an intimacy sprang up which doubtless had its effect on the education of young Williamson, for his father shared the opinions of the Mills in regard to religious, social and educational questions.

In or about 1840 Mr. Williamson retired from the India House on a pension. He soon afterwards gave up his house at Kensington and he and his family removed to the Continent and resided first in Paris and then near Dijon. At Dijon Alexander and his sister Antonia Helen, two years his senior, had private lessons. Concerning his pupils the tutor reported that the young lady had worked steadily, but of her brother so much could not be said. Later on Williamson spent a winter in Wiesbaden, working diligently at German, and he then went to Heidelberg to study medicine, in accordance with the wish of his father. He attended Tiedemann's lectures, but the professor was then very old, his lectures were uninteresting, and the students inattentive. On the other hand, Williamson found the lectures on chemistry by Gmelin very interesting, and notwithstanding his physical defects the work in the laboratory attracted him. The result was that he determined to be a chemist, his father's unwilling consent having been ultimately obtained.

In April, 1844, Williamson went to Giessen and entered the famous laboratory under Liebig. He lived for two years in the house of Professor Hillebrand, the professor of philosophy. In his first semester he attended, beside the lectures on chemistry, those of Bischoff on physiology. In writing to his father he spoke of the "clear and impressive" style of these lectures, which were given at seven o'clock in the morning. He seems to have been at first disappointed with Liebig's lectures, which he found "rather tedious from the extremely elementary manner in which he treated" the subject, and he was "not remarkably pleased with his delivery." However, in later letters he referred to Liebig's captivating manner and the expression of benevolence or of affection which a look of his could convey. And many years afterwards in his presidential address to the British Association, soon after the death of Liebig, he spoke with the greatest respect of the influence which had been exercised over his students by that great master.

Williamson seems to have been very industrious at Giessen, and he allowed himself no distraction from his work beyond such exercise, chiefly walking, as he found necessary for his health. Now and then he took part in a picnic or occasionally a dance. He testified to the general earnestness of the chemical students and their superiority as a class over the rest of the students at Giessen. His work at Giessen resulted in the publication of several papers, his first contributions to science. These appeared in the *Memoirs and Proceedings of the* (then young) *Chemical Society* of London.

A subject which seems to have occupied much of Williamson's thought during his first year at Giessen was the theory of galvanism, and in writing to his parents in 1845 he often refers to the experiments he is making in electricity. He even ventured to attack the theory of Humphry Davy, and though encouraged by Liebig, the professor of physics, Buff, did not attach so much importance to his views and the resulting paper was not published. In August, 1845, he received the degree of Ph.D.

From this time forward for three or four years Williamson partly suspended his chemical studies in order to devote himself to mathematics and physics. In the latter subject Buff gave him special facilities and allowed him access to the physical cabinet to which no student had been hitherto admitted, being only intended for the use of the lecturers.

In the summer of 1846 he went to Paris and placed himself under the tuition of Auguste Comte, who had been recommended by John Stuart Mill as the best man in Europe for completing a scientific education. He received a lesson in advanced mathematics three times a week and often spent his evenings at Comte's house with him and his disciples. The intimate association, during the three years he spent in Paris with such a man as Comte, must have had considerable influence on the development of his mind, and in after life Williamson showed the effect of this in the expression of his views and in occasional turns of phraseology. But he did not give up chemistry altogether, and set up a laboratory of his own in Paris at 8, Rue des Francs Bourgeois, where he carried on investigations, the results of which for the most part were not published but were probably preparatory to the views which a year or two later he expressed definitely on atomic movement and interchange.

Early in 1849 Professor Thomas Graham, of University College, London, made Williamson's acquaintance in Paris and encouraged him to become a candidate for the Professorship of Practical Chemistry in the College, which was then vacant by the death of its first occupant, George Fownes. Williamson applied and was appointed, and in the following October he began his long connection with the College and University which only came to an end in 1888, nearly forty years later. An account of his early work in this connection cannot be better given than in the words of Professor Carey Foster :

"At the beginning of his first college session Williamson read himself in by delivering a public introductory lecture to the courses of the Faculty of Arts and Laws. Of this lecture it is perhaps not unfair to say that the best part of it was the title, 'Development of Difference the Basis of Unity.' The discourse itself is disappointing and mostly consists of somewhat obvious generalities set forth with a philosophical air, but coming to no definite point. Graham kindly congratulated the lecturer on his musical voice."

Williamson's first few years at University College constituted a period of remarkable activity and productiveness. The first session of all produced his memorable paper on etherification, and it was soon followed by valuable contributions from his pupils ; among the earliest may be mentioned Duffy on Stearine and Wills on Heptylic Alcohol. The session 1853-54 was particularly fruitful, and furnished no less than six experimental papers by himself and his pupils. At this time he was a splendid teacher, always in the laboratory, going from one student to another, arousing and maintaining their interest in their work, and ready to discuss

compounds. What Williamson did was to show that ether was not derived from alcohol by the mere abstraction of water, but that the molecule of each contained the same amount of oxygen. The idea of classifying organic compounds according to types which in properties and reactions they were found to imitate had already been introduced; thus organic bases were all supposed to be derived from ammonia, which was taken as the *type*. In harmony with this idea Williamson introduced the use of the *water type*, and representing this compound as made up of two atoms of hydrogen to one atom of oxygen, he was able to bring into the same category not only inorganic acids, bases and salts, but alcohol and many organic acids and their salts.

But in his *Theory of the Formation of Æther* Williamson also introduced the idea of inter-molecular exchange and atomic motion, which lies at the foundation of the modern doctrine of chemical exchange and of ionisation.

In 1855 Graham, on being appointed Master of the Mint, resigned the chair of chemistry at University College, and Williamson was appointed to succeed him, retaining the professorship of analytical and practical chemistry. He entered on his new duties with enthusiasm, expending much time and money on the preparation and illustration of his lectures, in which he was for a time assisted by Henry Roscoe (afterwards Sir Henry), a former pupil. Recognition of the importance of his work led this year to his election into the Royal Society, and immediately after he had obtained his new position in the College he married Emma Catherine, third daughter of Thomas Hewitt Key, F.R.S., formerly professor of comparative grammar in the University of London and at this time headmaster of University College School.

The session ending with his appointment to the professorship of chemistry was the last in which Williamson published papers on the results of researches by himself or carried out at his instigation. However much regret may be felt on this decline of practical activity in the direction of chemistry, it is easy enough to find explanations and excuses. Williamson's own physical disabilities already referred to certainly interfered with the readiness with which laboratory operations could be undertaken by himself, but probably the chief reason for his neglect of chemistry was to be found in the attractions of a totally different field of work. From about 1854 and during several following years he became interested in the generation of steam for mechanical purposes, which he thought could be greatly improved, and he devised what he called a "tubulous" boiler which he ultimately patented. This invention, however, brought no profit to himself, and it is now uncertain to what extent it may have influenced inventors of a later day. A few years afterwards he also established at Willesden an experimental works where he made endeavours to improve certain manufacturing chemical processes. He was also interested in starting the Landore Steel Works, where Siemens' regenerative system of furnaces was put into operation.

Probably other influences were at work in diverting Williamson's attention from the subjects connected with his professorship. He was a man of restless activity of mind and looked for dominance in the various movements with which he became associated. His colleagues on the Council of the College learned to depend a good deal on his judgment in matters connected with administration and in the reconstitution of the College itself. Accordingly much of his time came to be absorbed in Committee work which would otherwise have been spent in the laboratory. He was also led to give much attention to the work of other bodies, such as the University of London, in which the institution of a Faculty of Science with degrees in the faculty was accomplished about this time. The work of the British Assocation and of the Royal and Chemical Societies also occupied much time. He served on the Council of the Royal Society in 1859-61 and in 1869-71; he became Foreign Secretary of the Society in 1873 and continued in that office for seventeen years.

Williamson gave several lectures to the Chemical Society in which he dealt with such subjects as valency, chemical nomenclature and especially the Atomic Theory of Dalton, of which he was a convinced exponent. It is interesting at the present time, fifty years later, to read the report of the discussion which followed this lecture, in which the leading English chemists of that time took part, and to observe the reluctance displayed by nearly all of them to accept the idea of atoms as physical units. It is instructive to compare the position of the theory at that time with its position at the present day, when not only is the idea of the particle accepted and justified by stereochemical facts and theories, but chemists and physicists are occupied in examining and discussing the constituent parts of the atom and in considering the various possible arrangements of these parts within,

In 1863 and 1864 Williamson became president of the Chemical Society and was elected to the same office in 1869 and 1870. He thus officiated as chairman on the occasion of the lecture in 1863 by Berthelot on synthesis of carbon compounds, and at the first Faraday Lecture in 1869, which was given by Dumas. The next year he inaugurated the important enterprise of issuing in the journal of the Society monthly reports in the form of abstracts of all papers dealing with chemical subjects which appeared in foreign or British publications. This system has been developed by many societies with great advantage to the spread of knowledge of the various subjects they represent.

Toward the end of 1863 Williamson received a number of young Japanese students, of whom three became boarders in his house. This was at the beginning of the period of change in Japan and the movement in favour of intercourse with Western forms of government and custom. These young men left their own country at the risk of their lives and some of them under assumed names. They came to England to make a systematic study of European science and civilisation, and no one could have been better qualified than Williamson to direct their studies and guide them in their observations, his own long familiarity with the life of France and Germany giving a broad outlook and freedom from insular prejudice. Most of these young Japanese, as well as their numerous successors, sent over by Prince Satsuma, afterwards attained distinguished positions in the government and administration of their own country. One of them, the Marquis Ito, was largely instrumental in framing the revised Japanese constitution.

From this time forward, however, the influence of Williamson on the progress of the science of chemistry appreciably declined. Whatever time he was able to secure from his multitudinous engagements outside appears to have been occupied with meditation on the results of the original and fertile views promulgated by himself in earlier days. He did not keep abreast of the most recent discoveries. The courses of practical chemistry at the College were left entirely to the care of assistants, with the result that the reputation of the place as a school of chemistry steadily dwindled. Most English chemical students of this time resorted to the College of Chemistry, where the professor (Hofmann) was daily in the laboratory and pursuing research. It was doubtless his influence which led so many of the students of that institution to proceed afterwards to Germany, where they obtained degrees by the results of work done in the University laboratories of that country. To such an extent was this customary that it began to be assumed that chemistry could be only effectively studied in Germany. For many years this idea was practically justified by the general inactivity of the British Universities. In spite of the gradual diversion of Williamson's attention from the practical work naturally associated with the duties of his chair, his students undoubtedly recognised his great originality of mind and his high qualities as a teacher. As a result of the demand for teaching in connection with chemical technology, a new professorship was established at University College. The old Birkbeck laboratory having been found not to provide the accommodation needed for the various classes, some new laboratories were put up under Williamson's direction in 1880. These have long since been superseded by the more modern and spacious buildings erected thirty years later under his successors.

In 1888 Williamson resigned his appointments in University College and was succeeded by William Ramsay. A good portrait of Williamson by the Hon. John Collier hangs in the Council Room at University College, and there is another painted by Biscombe Gardner in the Chemical Department of the College.

There are many of his former students and others surviving who remember clearly his tall, slight and upright figure, arrayed almost uniformly in grey trousers and frock-coat, hair and beard grey, with the peculiar look which distinguishes short-sighted people. Like other men of great originality of mind and force of character, he was wedded to his own views, even after they had long become unacceptable and practically inconvenient. He was doubtless right, for example, in introducing the revised nomenclature according to which sulphuric acid was called, for systematic purposes, *hydric* sulphate, but the consequence of retaining the name *acid* in accordance with previous custom derived from the time of Lavoisier for the oxides SO_3, CO_2, N_2O_5, etc., as he desired, would have been merely hopeless confusion, as it would have been practically impossible to extend the system to the numerous "organic" acids of which the anhydrides are unknown. There is no doubt Williamson was fond of power, and this perhaps swayed him to some extent in taking part in so many scientific and public bodies in which his influence was felt. On the other hand, there can be also no doubt that he was invariably filled with a high sense of duty, while the respect in which his memory is held by those who in the days of long ago were among his pupils is strong testimony to the benevolent influence he exercised over his juniors.

A considerable number of letters have been preserved from John Stuart Mill, Auguste Comte and other eminent persons, to Williamson in his young days and to his parents, beside many from Liebig and other of the most distinguished chemists at later periods. These present two points of interest. Some of them contain passages relating to current events, as, for example, a letter addressed by John Stuart Mill to A. W. Williamson's father on the subject of the French Revolution in 1848.

An amusing, if sad, comment on the way in which popular movements are brought about is afforded by an anecdote given at the end of one of Laurent's letters:

"Chez Madame Vve. Guaragnon:

"Pour quoi votre mari a-t-il voté? demandai-je à notre femme de ménage. Je n'en sais rien, me dit elle; il à dit *oui*, c'est tout. Alors il a voté pour Napoleon, lui dis-je. Il ne le sait pas, me répondit elle. On lui a dit, il faut écrire *oui*, comme tout le monde; sans cela tu n'auras pas d'ouvrage. Il a écrit *oui*, mais il ne sait pas ce que çela signifie."

He adds:

"Si le Père Éternel s'occupe de nous, il faut avouez qu'Il a quelquefois de singuliers caprices. Nous voilà Napoleonistes et par conséquence nous allons probablement redevenir les ennemis de la perfide Albion."

The majority of these letters were written to A. W. Williamson himself from scientific friends in France and Germany. There is one from Liebig congratulating him on his interesting research on ether and asking him to supply an account of it to the *Annalen*, always referred to as Liebig's, even down to the present day; another about the same time from Auguste Laurent referring to his inseparable colleague Gerhardt, to whom a monument in Strasbourg, long after his day, is now about to be erected. Laurent was in 1852 in a miserable state of health, and early in 1853 he died. There is also a letter from his widow acknowledging a gift of money from Williamson when she and her children were refugees in London after the disastrous defeat of France in 1870. At this time also there are several letters from Professor Berthelot showing that in the unhappy times when, as he says, "la folie sanguinaire de Napoleon III. nous eut jeté dans l'abime," the family escaped to England and received from Dr. and Mrs. Williamson "généreuse hospitalité." Berthelot had returned to France, and in May, 1871, was at Versailles while Paris was in the hands of the Commune.

Many letters from Professor Adolf Wurtz have been preserved, and among them one affording a kind of apology for the proposition at the beginning of his famous *History of Chemical Theory:* "Chemistry is a French Science. It was founded by Lavoisier, of immortal memory." Wurtz now writes: "Ne vous méprenez pas, cher ami, sur le sens de ma première phrase qui veut dire simplement que la chimie est née en France."

Another from the same friend in 1873 announces to Williamson his election as "Correspondant," that is, corresponding member of the French Academy of Sciences.

In the same year Williamson, as Foreign Secretary of the Royal Society, received from Professor Helmholtz, of Berlin, a letter which, in recognition of its great interest, is here transcribed. Helmholtz, previously Professor of Physiology at Heidelberg, was at this time Professor of Physics at the University of Berlin. He had given the fifth Faraday Lecture at the Royal Institution in 1881. The letter is as follows:

"Although I find myself unable for want of time and of health to come over to England, I did not wish to let pass the day when I should have received from the hands of your President the highly valuable token of benevolent approval which the Royal Society has deigned to confer on me [1] without repeating the expression of my gratitude to the Council and the Fellows of the Society. I keep firmly written in my memory the pleasant remembrance of former times when I was so happy as to assist at your meetings. I have felt deeply both the scientific weight of such a union of eminent men and the heartiness of the welcome with which the foreign guest was received. On these occasions I have learned to admire the organisation of this Society, kept together solely by the love of science and the voluntary exertions

[1] The Copley Medal.

of its fellows, governed solely by itself, the history of which, through a period of two hundred years, from Newton to Faraday, to the present moment, has been an uninterrupted sequence of glorious victories, which human intellect has won over the blind forces of Nature. I have seen in your country a good deal of strong enthusiasm and strong intellectual energy, devoted to scientific labour arising among men of different classes of society and of the most different occupations. I perceived that herein was the source of the strong individual originality, which is a characteristic of English science, and also the source of its practical fruitfulness. On the Continent the conditions of life for scientific men have been different; the greater part of them belonged always to a peculiar class, more isolated from other classes of men, more connected by its interests and its occupations. These conditions are more favourable to develop scientific schools, with all the advantages and all the disadvantages which tradition and the discipline of such a school tend to produce. Frenchmen turned more to the methodical and refined elaboration of detail. We Germans, partly driven by a native tendency, partly by the social and political consequences of our long religious struggles, turned more to the first principles of knowledge in general, and of scientific theories especially. I cannot disown this national tendency for myself; my own exertions have been devoted partly to the great natural law of the conservation of energy which lies at the root of all questions about the nature of force, partly to the physiological theory of nervous action and sensation which leads to that of perception, the source of all other knowledge. But I owe a great deal to England for my own intellectual education. Grown up among the traditions of high-flown metaphysics, I have learned to value the reality of facts in opposition to theoretical probabilities by the great example of English science. It was, for the greater part, this example which has sheltered me against losing myself in overstrained theoretical speculations.

"You will see from this confession that I have a personal right to give testimony to the good which results from international intercourse. But to acknowledge and esteem the superiority of others, to judge impartially how far it reaches, is not easy, where you don't meet with equal impartiality in return. Now here is another strong side of the Royal Society. It is a prerogative of great men and great natures that they are free to give acknowledgment without a trace of jealousy to whatsoever merits other men have attained. I wish that the Royal Society may continue to enjoy this prerogative, which ensures the highest and best fruits of international scientific intercourse."

The endeavour to promote and arrange international co-operation with the German and French chemical societies was an object which Williamson kept always in view, and testimony as to his activity in this direction is provided by letters received from Hofmann in Berlin as well as from Wurtz in Paris.

If any evidence were needed as to Williamson's position in the scientific world it would be found in this interesting collection of letters, from which it is clear that he was on terms of friendship with all the leading chemists of his time, including Dumas, Graham, Kekulé, Odling, Debus, Gerhardt and Laurent, Cannizzaro, Pasteur and many others, and that they seemed to turn to him for guidance on all sorts of occasions. He was also in communication with such influential people as George Grote, W. Stanley Jevons, James Prescott Joule, Clerk Maxwell and others.

Shortly before his retirement from the professorship Williamson built for himself a house at Hindhead, near Haslemere, and acquired land sufficient to afford scope for his interest in scientific farming. As he grew older his sight became more than ever imperfect and this apparently led to an accident in consequence of which he fell in the street and broke his arm. From this he recovered, but he died not long afterwards at his house, High Pitfold, Hindhead, on May 6, 1904, aged just eighty years. He was buried in the cemetery at Woking.

[Face p. 241]

Group X

CLASSIFICATION AND NATURE OF ELEMENTS

Mendeléeff (1834-1907) Crookes (1832-1919) Ramsay (1852-1916)

CHAPTER XIX

MENDELÉEFF [1]

To some few of the older among English chemists, the commanding, patriarchal figure of MENDELÉEFF was quite familiar. Though his several visits to London were often connected with official business of the Russian Government Department of Weights and Measures, of which he was the chief official during the later years of his life, he came several times with more purely scientific objects. In 1889 the occasion of his presence in London was the Faraday Lecture which he had been invited to give to the Chemical Society, but which, owing to a sudden and urgent recall to his home, he was unable to deliver in person. His last appearance in this country was in November, 1905, when the Copley Medal was awarded to him by the Royal Society.

His name, more than any other, will be for ever associated with the development of the great generalisation known as the periodic system of the elements.

Dmitri Ivanovitsch Mendéléeff [2] was the fourteenth and youngest child of his parents, Ivan Pavlovitsch and Maria Dmitrievna, *née* Kornileff. His father, a former student of the Chief Pedagogic Institute of St. Petersburg, obtained the

[1] The following account is adapted from the Mendeléeff Memorial Lecture given to the Chemical Society, October 21, 1909, by the author.

[2] For many of the details of Mendeléeff's career and of his home life the writer is indebted to the family chronicle compiled, soon after his death, by his niece, N. J. Gubkina (*née* Kapustina), and published in St. Petersburg, also to pamphlets by A. Archangelsky and P. J. Robinowitsch. He also desires to express his thanks to Mr. D. V. Jéquier, formerly of Petrograd, as well as to several Russian friends, for indispensable assistance in translation.

appointment of Director of the Gymnasium at Tobolsk, in Siberia, where he met Maria Dmitrievna, who became his wife. After a few years at Tobolsk, he was transferred to school directorships in Russia, first at Tambov, and afterwards at Saratov. But in order to satisfy the ardent wish of his wife, he took advantage of an opportunity of exchange, by which he became once more Director of the College at Tobolsk, and the family returned to Siberia. Here on January 27, 1834 (O.S.) was born Dmitri Ivanovitsch, the youngest son. Soon after his birth the father became gradually blind from cataract in both eyes, and was obliged to resign, the whole family, including eight children, having to subsist on a small pension of 1,000 roubles (about £100 per annum). The mother, Maria Dmitrievna, belonged to the old Russian family, Kornileff, settled at Tobolsk. They were the first to establish in Siberia the manufacture of paper and glass. In 1787 the grandfather of Dmitri opened at Tobolsk the first printing press, and in 1789 produced the first newspaper in Siberia, the *Irtysch*. The glass works were situated in the village of Aremziansk, a short distance from Tobolsk.

According to the family tradition, one of the Kornileffs in a previous generation had married a Khirgis Tartar beauty, whom he loved so passionately that when she died he also died of grief. The pure Russian blood thus received a strain of the Mongolian race, and some of their descendants preserved traces of the Oriental type. This, however, was not very noticeable in the features of the chemist.

From her childhood, Maria Dmitrievna was distinguished by her intelligent wish for instruction, and having no other resource when her brother Basile went to school she repeated by herself all his lessons, and thus, unaided, obtained some part of the knowledge so eagerly desired. There can be no doubt she was a woman possessed of remarkable vigour of mind, who exercised great influence over her children. Her activity and capacity are further illustrated by the fact that when her husband became blind she revived the business of the glass works, and carried it on till after his death from consumption in 1847.

Tobolsk was at that time a place of banishment for many political exiles, the so-called Decembrists, one of whom, Bassargin, married Olga, an elder sister of Dmitri. To these Decembrists the boy owed his first interest in natural science. His mother had always cherished the hope that at least one of her children would devote himself to science, and accordingly, after her husband's death and the destruction of the glass works by fire, and spite of failing health and scanty means, she undertook the long and tedious journey from Tobolsk to Moscow, accompanied by her remaining children, Elizabeth and Dmitri Ivanovitsch, with the object of entering the latter, then nearly fifteen years of age, at the University. Disappointed in this object, owing to official difficulties, she removed in the spring of 1850 to St. Petersburg, where ultimately, with the assistance of the Director, Pletnoff, of the Central Pedagogic Institute, a friend of her late husband, she succeeded in securing for her son admission to the Physico-Mathematical Faculty of the Institute, together with much-needed pecuniary assistance from the Government.

The debt which Dmitri Ivanovitsch owed his mother he acknowledged later in the introduction to his work on *Solutions*, which he dedicated to her memory in the following interesting lines:

"This investigation is dedicated to the memory of a mother by her youngest offspring. Conducting a factory, she could educate him only by her own work. She instructed by example, corrected with love, and in order to devote him to science she left Siberia with him, spending thus her last resources and strength. When dying, she said, 'Refrain from illusions, insist on work, and not on words. Patiently search divine and scientific truth.' She understood how often dialectical methods deceive, how much there is still to be learned, and how, with the aid of science, without violence, with love but firmness, all superstition, untruth, and error are removed, bringing in their stead the safety of discovered truth, freedom for further development, general welfare, and inward happiness. Dmitri Mendeléeff regards as sacred a mother's dying words. October, 1887."

In the Pedagogic Institute Dmitri Ivanovitsch was thus able to devote himself to the mathematical and physical sciences under the guidance of Professors Leng and Kupfer in physics, Woskresensky in chemistry, and Ostragradsky in mathematics. Unfortunately, at the end of his course, his health failed, and about this time his mother died. Having been ordered to the South, he fortunately obtained an appointment as chief science master at Simferopol, in the Crimea. The southern climate soon alleviated the serious symptoms of lung disorder, and removal being necessary in consequence of the Crimean War, he was able soon afterwards to undertake a post as teacher of mathematics and physics at the Gymnasium at Odessa. In 1856 he returned to St. Petersburg, and at the early age of twenty-two he was appointed *privat docent* in the University, having secured his certificate as master in chemistry.

At this time he appears to have passed rapidly from one subject to another, but he soon found matter for serious and protracted study in the physical properties of liquids, especially in their expansion by heat. And when, in 1859, by permission of the Minister of Public Instruction, Mendeléeff proceeded tostudy under Regnault in Paris and afterwards in Heidelberg, he devoted himself to this work, communicating his results to Liebig's *Annalen* and the French Academy of Sciences. Returning two years later to St. Petersburg, he secured his Doctorate, and was soon afterwards appointed Professor of Chemistry in the Technological Institute. In 1866 he became Professor of General Chemistry in the University, Butlerow at the same time occupying the Chair of Organic Chemistry.

As a teacher, Mendeléeff seems to have possessed a special talent for rousing a desire for knowledge, and his lecture room was often filled with students from all faculties of the University. Many of his former students remember gratefully the influence he exercised over them.[1] One of these writes : "I was a student in the Technological Institute from 1867 to 1869. Mendeléeff was our professor, and in 1868 taught organic chemistry. The previous course by the professor of inorganic chemistry consisted of a collection of recipes, very hard to remember, but, thanks to Mendeléeff, I began to perceive that chemistry was really a science. The most remarkable thing at his lectures was that the mind of his audience worked with his, foreseeing the conclusions he might arrive at, and feeling happy when he did reach these conclusions. More than once he said, 'I do not wish to cram you with facts, but I want you to be able to read chemical treatises and other literature, to be able to analyse them, and, in fact, to understand chemistry. And you should remember

---

[1] For the following reminiscences, the writer is indebted to Mr. L. Goldenberg and Prince P. Kropotkin respectively.

that hypotheses are not theories. By a theory I mean a conclusion drawn from the accumulated facts we now possess which enables us to foresee new facts which we do not yet know.' He was considered among the students a most liberal man, and they thought of him as a comrade. More than once during a disturbance between the students and the administration Mendeléeff supported the students, and under his influence many matters were put right." (L. G.) Another former student in the University writes as follows: "I am sorry to say I did not know Mendeléeff personally. I only had the good fortune to follow, in the years 1867-69, his lectures on both Organic and Inorganic Chemistry. The former was an abridged course, which he had the admirable idea to deliver for us students of the mathematical branch of the physico-mathematical faculty. He reduced this course of one lecture a week during one year to a general review of organic compounds and the general laws of their structure. You can imagine what it must have been in the hands of Mendeléeff, thirty-three or thirty-four years old at that time, in the full enjoyment of his mental powers, and just then plunged into the study of his great generalisations. For me it was a revelation, being occupied with the great questions connected with the development of the new system of atomic weights, the mechanical theory of heat, etc. Grove's, Thomson's, Joule's, Séguin's works were then just out, and in these years a sudden blossoming of the natural sciences in all directions seemed to bring us near to the solution of the great problems of the nature of matter and of gravitation. Then I followed Mendeléeff's lectures on Inorganic Chemistry. The *Principles of Chemistry* was not yet out, but he was evidently writing it at that time. You know how much is said in the footnotes to his *Principles*; well, imagine each of these notes developed into a beautiful improvisation, with all the freshness of thought of a man who, while he speaks, evolves all the arguments for and against, there on the spot. The hall was always crowded with something like two hundred students, many of whom, I am afraid, could not follow Mendeléeff, but for the few of us who could it was a stimulant to the intellect and a lesson in scientific thinking which must have left deep traces in their development, as it did in mine." (P. K.)

One of Mendeléeff's most remarkable personal features was his flowing abundance of hair. The story goes that, before he was presented to the Emperor, Alexander III., his Majesty was curious to know whether the professor would have his hair cut. This, however, was not done, and he appeared at Court without passing under the hands of the barber. His habit was to cut his hair once a year in spring, before the warm weather set in. His eyes, though rather deep set, were bright blue, and to the end of his life retained their penetrating glance. Tall in stature, though with slightly stooping shoulders, his hands noticeable for their fine form and expressive gestures, the whole figure proclaimed the grand Russian of the province of Tver.

At home, Mendeléeff always wore an easy garment of his own design, something like a Norfolk jacket without a belt, of dark grey cloth. He rarely wore uniform or evening coat, and attached no importance to ribbons and decorations, of which he had many.

As to his views on social and political questions, many people thought him a rigid monarchist, but he said of himself that he was an evolutionist of peaceable type, desiring a new religion, of which the characteristic should be subordination of the individual to the general good. He always viewed with much sympathy what is called the feminine question. At the Office of Weights and Measures he employed several ladies, and about 1870 he gave lectures on chemistry to classes of ladies. Nevertheless he considered women inferior to men both in business and in intellectual pursuits, and he thought the chief promoters of the feminine movement aimed, not so much at equality of political position, as at opportunities for work and to escape inactivity. But he thought the feminine temperament specially suited to all branches of art in the broadest sense of the word, including education.

Mendeléeff held decided views on the subject of education, which he set forth in several publications, especially *Remarks on Public Instruction in Russia* (1901). Here he says: "The fundamental direction of Russian education should be living and real, not based on dead languages, grammatical rules, and dialectical discussions, which, without experimental control, bring self-deceit, illusion, presumption and selfishness." Believing in the soothing effect of a vital realism in schools, he considered that universal peace and the brotherhood of nations could only be brought about by the operation of this principle. Speaking of the reforms desirable, he says that "for such reforms are required many strong realists; classicists are only fit to be landowners, capitalists, civil servants, men of letters, critics, describing and discussing, but helping only indirectly the cause of popular needs. We could live at the present day without a Plato, but a double number of Newtons is required to discover the secrets of nature, and to bring life into harmony with the laws of nature." Mendeléeff was evidently a philosopher of the same type as our own Francis Bacon.

"I am not afraid," he says later, "of the admission of foreign, even of socialistic, ideas into Russia, because I have faith in the Russian people, who have already got rid of the Tartar domination and the feudal system."

Mendeléeff always dined at six o'clock, and liked to entertain his friends and relations, but in his own diet he was extremely moderate. After dinner he enjoyed reading light literature, especially books of adventure, such as those of Fenimore Cooper or Jules Verne. But his literary tastes were peculiar. Though interested in serious literature and appreciating Shakespeare, Schiller, Goethe, Victor Hugo, and Byron as well as the Russian classics, beginning with Zhoucovsky and Pouschkin, his favourites among Russian poets were Maïcoff and Tuttcheff, and among the rest Byron. Of the last-named he preferred to all his other works the gloomy poem called "Darkness," and among the rest the "Silentium" of Tuttcheff.

He rarely went to the theatre, and did not approve of frequent visits to the theatre by his children, as he considered such distractions tend to destroy concentration and fill the mind with "trifles and foolishness." On the other hand, he was very fond of pictures, and he visited all the exhibitions. That he was interested in questions relating to art, and had given much thought to æsthetic problems, is indicated by a letter [1] which he addressed in November, 1880, to the well-known Russian daily paper at that time, *Goloss* (*The Voice*), on the subject of a picture by Kouindji, "Night in the Ukraine." Writing of the influence of landscape on different minds, he says: "At first it seemed to me a matter of personal taste, of individual sensitiveness of different persons to the beauty of nature." But, rejecting this simple view, he was led to a conception which he regarded as really satisfactory, and which he wished to share with others. He says: "Landscape was depicted in antiquity, but was not in favour in those times. Even the great masters of the sixteenth century made use of it merely as a frame to their pictures.

[1] Considerably condensed in the following abstract.

It was the human form which principally inspired artists of that epoch; even the gods and the Almighty Himself appeared to their minds in human shape. In this alone they found the infinite, the inspiring, the divine. And this was because they worshipped human mind and human spirit. This found expression in science in an exceptional development of mathematics, logic, metaphysics and politics. Later, however, men lost faith in the absolute and original power of human reason, and they discovered that the study of external nature assists even in the correct appreciation of the nature of the human inner self. Thus nature became an object of study; a natural science arose unknown either to antiquity or to the period of the Renaissance. Observation and experience, inductive reasoning, submission to the inevitable, soon gave rise to a new and more powerful, more productive method of seeking truth. It thus became evident that human nature, including its consciousness and reason, is merely a part of the whole, which is easier to comprehend as such from the study of external nature than of the inner man. External nature thus ceased to be merely subservient to man, and became his equal, his friend. Dead and senseless as it had been, it now became alive. Everywhere it presented motion, stores of energy, natural reason, simplicity and plan. Inductive and experimental science became a crown of knowledge, royal metaphysics and mathematics had now to be content with modest questioning of nature. Landscape painting was born simultaneously with this change, or perhaps a little earlier. Thus it will probably come to pass that our age will hereafter be known as the epoch of natural science in philosophy, and of landscape in art. Both derive their materials from sources external to man. . . . Man has, however, not been lost sight of as an object of study and of artistic creation, but he now appears, not as a potentate or as a microcosm, but merely as part of a complex whole."

In 1863, when twenty-nine years of age, Mendeléeff married his first wife, *née* Lestshoff, by whom he had one son, Vladimir,[1] and a daughter, Olga; but the marriage proved unhappy, and after living apart for some time there was a divorce. In 1877 he fell in love with a young lady artist, Anna Ivanovna Popova, of Cossack origin, and in 1881 they were married. This lady

[1] Died in 1899, aged 34.

exercised considerable influence over his views about art, and the walls of his study were furnished with many products of her pencil, notably portraits of Lavoisier, Descartes, Newton, Galileo, Copernicus, Graham, Mitscherlich, Rose, Chevreul, Faraday, Berthelot, and Dumas, and others of relatives. After his second marriage Mendeléeff lived first at the University, and afterwards in the apartments built for the Director at the Bureau of Weights and Measures, and here his younger children were born, Lioubov (Aimée), Ivan (Jean), and the twins, Maria and Vassili (Basile).

In 1890, in consequence of a difference with the administration, Mendeléeff retired from the Professorship in the University. During the disturbances among the students in that year, he succeeded in pacifying them by promising to present their petition to the Minister of Education. Instead of thanks for this service, however, the Professor received a sharp reprimand from the authorities for not minding his own business. The consequence was that Mendeléeff resigned. Independently of the petition, however, there were probably deeper reasons for his being out of favour with the Ministry, connected with his irreconcilable enmity to the classical system of education already referred to. Of this he had made no secret, and it had already brought him into conflict with the authorities. In 1893, however, he was appointed by M. Witte to the office of Director of the Bureau of Weights and Measures, which he retained till his death.

In the earlier part of his life, Mendeléeff was interested in carrying on a series of agricultural experiments on his Tver estate, Boblovo. The peasants, much struck by his success and the abundance of his crops, inquired of him whether this was due to his luck or to his "talent." With a smile and the patois which he always affected in speaking to the country people, he informed them that he certainly had "talent," and, as he said afterwards at home, there is no merit in having luck.

Once during the solar eclipse in 1887 he ascended alone in a balloon with the object of making scientific observations. His assistant, Kovanko, who sat with him in the basket, alighted at the last moment, probably ordered to do so by his chief because the balloon would not rise. When the balloon shot up quickly and disappeared in the clouds, his family was naturally very much alarmed. Fortunately the hero of the adventure was able to descend safely, and a few hours later returned to his family from

Moscow. The peasant women thereafter used to tell that Dmitri Ivanovitsch flew on a bubble and pierced the sky, and for this the authorities made him a chemist!

Mendeléeff was very democratic in his habits, and when travelling from the Capital to his estate, six or seven hours by rail, he always made use of the third class, and on the way talked freely to his fellow-passengers on a variety of subjects, so that at the end of the journey he was surrounded by all sorts of people. At the railway station, about twelve miles from Boblovo, he was always met by the same driver, Zassorin, who with his troika of greys transported the whole family at full gallop, according to Russian custom.

Such, then, are the chief features of a great personality. If it be admitted that stories are told of his occasional irritability of temper, we can well place on the other side of the account the cordial relations always subsisting between the Professor and his assistants, the confidence and respect between the Master and his servants, the deep affection between the Father and his children, which are known to have persisted throughout his life, and which could be illustrated by many anecdotes. These stories merely serve "to give the world assurance of a man."

For us who live on the other side of Europe, separated as we are by race, by language, by national and social customs, and by form of government, it is not easy to understand completely the texture of such a mind, the quality of such genius, and the conditions, social or political, which may have served to encourage or to repress its activity. The Russian language may be eloquent, expressive, versatile and harmonious, or it may possess any other good quality that may be claimed for it by those to whom it is a mother tongue, but the fact remains that it is a barrier to free intercourse between the Russian people and the world outside their country. This alone creates a condition which must influence the development of thought, and must give to Russian science and philosophy a colour of its own. Mendeléeff was, like many educated Russians, a man of very liberal views on such subjects as education, the position of women, on art and science, and probably on national government. We can hardly guess what would be the influence on such a nature of a rigid administrative *régime* which forbade even the discussion of such questions. We in England are almost unable to imagine such a state of things as would be represented by the closing of, say, University

College for a year or more, because the question whether the House of Lords ought to be abolished had been debated in the Students' Union. Imagine the Professor of Chemistry, along with his colleagues, for such a reason deprived of the use of his laboratory by the police, and only allowed to resume his studies when someone down at Scotland Yard thought proper. Such being the experience of most of the Russian Universities and Technical High Schools, it is not surprising that the output of Russian science, notwithstanding the acknowledged genius of the Russian people, appears sometimes comparatively small. The amount of work done by Mendeléeff, both experimental and theoretical, was prodigious, and all the more remarkable considering the cloudy atmosphere under which so much of it was accomplished. Unfavourable as were the conditions under which he did his work, they did at any rate not altogether prevent his accomplishing much that was valuable,[1] and it may be regarded as fortunate that it was completed before the revolution which has swept out of his unhappy country most of the accumulated treasures of civilisation and for the present has destroyed science, letters and arts, as well as all who made the cultivation of intellectual pursuits the business of their lives. So far as is known the Academy of Sciences has gone down in the general ruin.

In 1882 the Royal Society conferred on Mendeléeff, jointly with Lothar Meyer, the Davy Medal. In 1883 the Chemical Society elected him an Honorary Member, and in 1889 it conferred upon him the highest distinction in its power to award, namely, the Faraday Lectureship, with which is associated the Faraday Medal. In 1890 he was elected a Foreign Member of the Royal Society, and in 1905 he received the Copley Medal. So far as England is concerned, his services to science received full acknowledgment. It is all the more remarkable, therefore, that he never became a member of the Imperial Academy of Sciences of St. Petersburg.

Towards the end of 1906 Mendeléeff's health began to fail. Nevertheless he was able to attend the Minister on the occasion

[1] Professor Walden, at the end of a biographical notice published in the *Berichte d. Deut. Chem. Ges.*, April, 1909, gives a list of 262 printed publications by Mendeléeff. These include, not only memoirs on physical and chemical subjects, but books, pamphlets, reports, and newspaper articles relating to exhibitions, to the industries of Russia, to weights and measures, to education, to art, and even to spiritualism.

of an official visit in January to the office of Weights and Measures, but he caught cold and, enfeebled as he had been by influenza in the preceding autumn, inflammation of the lungs set in. Retaining consciousness almost to the last, he requested even on the day of his death to be read to from the *Journey to the North Pole*, by his favourite author, Jules Verne. He died in the early morning of January 20 (O.S.), 1907, within a few days of his seventy-third birthday. He was buried in the Wolkowo Cemetery beside the graves of his mother and son.

Turning now to a survey of Mendeléeff's work as a man of science, it will be sufficient if we pass lightly over his first essays. Like so many other chemists, he began by handling simple questions of fact, his first paper, dated 1854, when he was twenty years of age, being on the composition of certain specimens of orthite. It was not till 1859 that he settled down to serious examination of the physical properties of liquids, which led him to a long series of experiments on the thermal dilatation of liquids, of which the chief ultimate outcome was the establishment of a simple expression for the expansion of liquids between 0° and the boiling point. (Chem. Soc. Trans., 1883, **45**, p. 126). This formula is liable to the same kind of modification which has been found necessary in the case of gases. It is, of course, applicable only to an ideal liquid from which all known liquids differ by reason of differences of chemical constitution and consequent differences of density, viscosity, and other properties. Thorpe and Rücker, by applying van der Waals' theory of the general relation between the pressure, volume, and temperature of bodies to Mendeléeff's expression for the thermal expansion, developed a simple method of calculating the critical temperature of liquids from observations of their expansion (Chem. Soc. Trans., 1884, **45** p. 135).

Mendeléeff devoted a large amount of time and of experimental skill to the estimation of the densities of various solutions, especially mixtures of alcohol and water and of sulphuric acid and water, and of aqueous solutions of a large number of salts. In 1889 he embodied the whole in the monograph already referred to. In a paper communicated to the Transactions in 1887 (**51**, p. 779), he stated his views in the following words: "Solutions may be regarded as strictly definite atomic chemical combinations at temperatures higher than their dissociation temperatures. Definite chemical substances may be either formed or decomposed at temperatures which are higher than those at which dissociation commences; the same phenomenon occurs in solutions; at ordinary temperatures they can be either formed or decomposed."

In conjunction with some of his students, Mendeléeff also studied minutely the question of the elasticity of gases, and published several papers on the subject, extending over a period of some ten years from 1872.

Another subject to which Mendeléeff gave a good deal of attention was the nature and origin of petroleum. Having already reported in 1866 on the naphtha springs in the Caucasus, in the summer of 1876 he crossed the Atlantic and surveyed the oil fields of Pennsylvania. In the course of these investigations, he was led to form a new theory of the mode of production of these natural deposits. The assumption that the oil is a product of the decomposition of organic remains he rejects on a variety of grounds, which are set forth in a communication to the Russian Chemical Society. (Abstract, see *Ber.*, 1877, **10**, p. 229.) Mendeléeff assumes, as others have done, that the interior of the earth consists largely of carbides of ametals, especially iron, and that hydrocarbons result from the penetration of water into contact with these compounds, ametallic oxide being formed simultaneously. The hydrocarbons are supposed to be driven in vapour from the lower strata, where temperature is high, to more superficial strata, where they condense and are retained under pressure. In 1886, in consequence of rumours as to the possible exhaustion of the Russian oil fields, he was sent by the Government to Baku to collect information, and in 1889 he made a communication on this subject to Dr. Ludwig Mond, which is printed in the *Journal of the Society of Chemical Industry* (1889, **8**, p. 753).

The great generalisation known as the Periodic Law was not arrived at by one step, nor by the ideas of any one man. No general scheme of atomic weights had been possible, partial and imperfect efforts in this direction being represented by Dœbereiner's triads and the application of the principle of homology made use of by Dumas. The first step was taken by J. A. R. Newlands, who, after some preliminary attempts in 1864-5, discovered that when the elements are placed in the order of the numerical value of their atomic weights, corrected as advised by Cannizzaro, the eighth element starting from any point on the list exhibits a revival of the characteristics of the first. This undoubtedly represents the first recognition of the principle of periodicity in the series of atomic weights, but whether discouraged by the cool reception of his "law of octaves" by the chemical world or from imperfect apprehension of the importance of this discovery, Newlands failed to follow up the inquiry. It was not long, however, before the matter was taken up by others. Odling especially seems to have given much thought to the subject, and, ignoring Newlands' previous attempts, he drew up towards the end of 1864 a table containing a list of all the then well-known elements, arranged horizontally in the order of their generally accepted groups, and perpendicularly in the order of their several atomic weights. He concluded an article in Watts's Dictionary a few months later with these words: "Doubtless some of the arithmetical relations exemplified in the foregoing table are merely accidental, but, taken altogether, they are too numerous and decided not to depend on some *hitherto unrecognised* law."

Such, then, was the state of knowledge about this time.

In March, 1869, Mendeléeff communicated to the Russian Chemical Society an enunciation of the principle of periodicity and a statement of some of the consequences of this recognition of the relation of properties to atomic weight throughout the whole range of the known elements, and this statement was accompanied by a table which, while it bears a close resemblance to Odling's table of 1864, was apparently connected in his mind with an idea which became clearer and more decisive in the modifications which he immediately afterwards introduced into the arrangement.

Mendeléeff's table of 1869 was in 1871 modified so as to assume the form with which we have all been so long familiar, and which is to be found in every modern text-book. Thus it may be claimed for Mendeléeff that he was actually the first, not only to formulate a general law connecting atomic weights with properties, but was the first to indicate its character, and, as himself (*Principles*, 1905, II, p. 28) has pointed out, he was the first "to foretell the *properties of undiscovered* elements, or to alter the accepted atomic weights" in confidence of its validity. The time was, in fact, ripe for the enunciation of this general principle, and, the suggestion once given, the relations embodied in the law could not fail to attract other chemists. Accordingly, in December, 1869, Lothar Meyer, with such knowledge of Mendeléeff's scheme as could be derived from the imperfect German version of his paper of the previous March, proved himself a convinced exponent of the idea by contributing to Liebig's *Annalen* a paper containing a table, substantially identical with that of Mendeléeff, and diagram of atomic volumes, which, more clearly even than the tabular scheme, illustrated the principle of periodicity.

The history of science shows many instances of the same kind. Great generalisations have often resulted from the gradual accumulation of facts which, after remaining for a time isolated or confused, have been found to admit of co-ordination into a comprehensive scheme, and, this once clearly formulated, many workers are found ready to assist in its development. The case is nearly parallel to the recognition of the operation of natural selection by Darwin and Wallace, or it might be compared to the discovery of oxygen by Priestley and Scheele and the utilisation of this knowledge by Lavoisier. In each case much preparatory work had been done, and a body of knowledge had been gradually accumulated which, when duly marshalled and surveyed by the eye of a master, could scarcely fail to reveal to him the underlying principle. The full consequences, however, would appear only to a few.

The law of periodicity was expressed by Mendeléeff in the following words:[1]

"The properties of the elements, as well as the forms and properties of their compounds, are in periodic dependence on, or (expressing ourselves algebraically) form a periodic function of, the atomic weights of the elements." After a brief historical account of the discovery of the law by himself, Mendeléeff concludes by saying (*Principles*, p. 18): "I consider it well to observe that no law of nature, however general, has been established all at once; its recognition has always been preceded by many presentiments; the establishment of a law, however, does not take place when the first thought of it takes form, or even when its significance is recognised, but only when it has been confirmed by the results of experiment which the man of science must consider as the only proof of the correctness of his conjectures and opinions."

[1] *Principles*, 1905, vol. ii., p. 17.

The confirmation in this case was provided by the remarkable prediction by Mendeléeff of the properties of elements then unknown, subsequently identified with scandium, gallium, and germanium.

The system once accepted could be applied to such purposes as the correction of atomic weights, illustrated by the case of beryllium, the recognition of previously unnoticed relations, and the discovery of new elements, notably the companions of argon by Ramsay some years later.

One result of the recognition of the periodic law is that theories concerning the genesis of the elements received a stimulus previously unknown. It is, however, interesting to note the attitude of Mendeléeff toward this question, and the small extent to which this attitude appears to have become modified with the lapse of time. When, in 1889, twenty years after the discovery of the law, he composed the Faraday lecture, he seems to have regarded speculation in this direction as a kind of abuse of the periodic system, and attempts in this direction were classed by him among mere utopias.

Chemists and physicists have, however, found it impossible to resist the fascination of this problem, and accordingly there have been many hypotheses as to the origin of the elements and the nature of their connection with one another. These seem to be inseparable from the periodic scheme itself, which at once provokes the inquiry, Why do these numerical relations occur, and what is the meaning of them if they do not point to a common genesis or the operation of some process of evolution?

Hypotheses concerning the evolution of the elements were formerly usually based on the assumption that the successive stages of condensation of elemental matter proceeded from a single primary stuff, which, by a process analogous to polymerisation among carbon compounds, gave rise to atoms of greater and greater mass, which were stable at the prevailing and any lower temperature. The physical cause of the successive condensations is supposed to be a falling temperature. It is, of course, possible to imagine that if to the stuff of which hydrogen atoms consist are added successive portions of matter of the same kind, stable structures may at intervals result which we know as the atoms of the elements helium, lithium, beryllium, boron, carbon, nitrogen, oxygen, and fluorine, provided the idea of internal structure in these atoms is allowed. Otherwise, from the mere accretion of matter upon a central nucleus, there seems no sufficient reason why there should not have been formed an indefinite number of intermediate masses corresponding to an indefinite number of what would be called elements. Further, it is difficult to understand why simple increase of mass should change, say, oxygen into fluorine, while a further addition of the same kind should change negative fluorine into inert neon or positive sodium. The possibility of the condensation of a single "protyl" so as to produce, at successive though unequal stages of cooling, the elements known to the chemist was discussed long ago by Crookes.

This hypothesis, however, was put forward long before the work of Sir J. J. Thomson and his school was given to the world and the electron was accepted as a physical reality. The hypothesis then arose that one elemental stuff may give rise to the whole array of known elements by a process of condensation accompanied by a loss or gain of electrons, the mass of each of which is approximately one-eighteen-hundredth of the mass of an atom of hydrogen.

If we assume that atoms are made up of two parts (protyls), positive and negative, in proportions which determine by the preponderance of one or the other whether the element shall exhibit the positive character of a metal like lithium or the negative character of a halogen, we arrive at a hypothesis which recalls the ideas put forward nearly a century ago by Berzelius. His views are familiar to every student of the history of chemistry, but have long been relegated to the lumber room of worn-out doctrine. The last few years have, however, given us the remarkable experimental investigations of J. J. Thomson already referred to, and the new conceptions concerning the nature of atoms, which revive the fundamental idea that they are made up of two components.

The conceptions presented to us in J. J. Thomson's work permit of several supplementary hypotheses, especially the idea that if atoms are really made up of smaller corpuscles these are not thrown together in confusion but must be distributed within the mass in a definite order, which is determined by the attraction of the electro-positive nucleus and the self-repulsion of the negative corpuscles included in it. Once the idea of structure within the atom is admitted, the possibility presents itself of there being for the same mass more than one arrangement corresponding to what is called isomerism in compounds. In this way the case of elements with similar properties and identical or nearly identical atomic weights may be explained. It does not seem very long ago in the memory of many now living that the nature of the isomerism of the derivatives of benzene was a deep mystery, from which nearly all obscurity was cleared away in the light of the then new theory of the constitution of benzene. Something of the kind may yet be found in the problem of the elements.

Any account of Mendeléeff's work must contain a reference to his remarkable book, *The Principles of Chemistry*, which in its day had an immense influence in the promulgation of his views concerning the elements and their relations to one another. Several English editions appeared, but though it must be regarded as a work of genius it was never very well adapted to the use of beginners, and the advance of knowledge has now rendered it obsolete. In the interesting preface to the book Mendeléeff pointed out how the peculiarities of a book represent a reflection of the personality of the author. In the present case author and book are now alike things of the past, but much of their influence remains.

The views of Boyle, of Lavoisier and of Dalton have been corrected by experience and broadened by extended knowledge, but the fundamental and essential parts of their ideas remain, and their names are immortal. In like manner the expression of the Periodic Law of the elements as known to the present generation has already been absorbed into a more comprehensive scheme, and doubt as to the doctrine of evolution has been practically resolved. But as with the Atomic Theory itself, there is no reason to doubt that the essential features of the periodic scheme will be clearly distinguished through all time.

William Crookes.

[Face p. 259.]

CHAPTER XX

CROOKES

THE investigation which led Williamson to his theory of the formation of ether and seemed to justify his hypothesis of inter-atomic movement and interchange was one of the first in the domain which now constitutes an entire department of the science under the name *physical chemistry*. The physical properties of the elements and their compounds have long been a subject of enquiry, but the study of the conditions under which chemical combinations and decompositions proceed has been the work of many men and many years in more recent times. Among the phenomena which during the last fifty years have attracted much attention and the study of which has led in very recent times to the most surprising results, are those which are observed when a discharge of electricity occurs in a gas in a more or less expanded state. This is a case which illustrates in a striking way the dependence of discovery on the invention of new apparatus. This is especially true of the apparatus now available for the production of what is called a *vacuum*, beginning with the mercury pump invented by Hermann Sprengel in 1865. The development of apparatus for the production of high potential discharge has also been an important feature of the most modern work. The late Sir William Crookes must be regarded as the chief pioneer in the investigation of phenomena shown in gases under greatly reduced pressure, and his researches must be viewed as the starting-point for the discoveries by Sir J. J. Thomson and his school, which have thrown a new light on the constitution of matter. The following account of his life and work is taken to a large extent from the obituary notice prepared by the author at the request of the Council for the *Proceedings of the Royal Society*.

William Crookes was born in London, June 17, 1832. His father, Joseph Crookes, born in 1792, was the son of a small tailor in the north of England. He came to London early in the nineteenth century to seek his fortune, and he ultimately established a tailor's business in Regent Street which proved very prosperous. He died at Brook Green, Hammersmith, in 1884 at the age of ninety-two. Joseph Crookes married at Aynhoe, Northamptonshire, as his second wife, Mary Scott, and from this lady he had a second family of several sons and daughters. William was her first-born, and according to those who remember his mother, William resembled her in feature and in disposition.

There is but little on record concerning his earliest years, but he used to maintain that, improbable as it may appear, he remembered learning to stand and to walk. Such systematic education as he received was obtained at a grammar school at Chippenham. At home his inclination toward experiment was manifested in various ways, and ultimately his father yielded to his expressed wish and allowed him to enter, at the early age of sixteen, the then newly-established Royal College of Chemistry. This occupied a building specially erected in 1846 on the south side of Oxford Street, near to the top of Regent Street, and now with the addition of a storey providing offices for the British Medical Council.

Crookes' first paper resulting from experimental research on the selenocyanides was published in 1851. He therefore must have been both an apt and diligent student of chemistry from the first, and the recognition of these qualities by Professor Hofmann no doubt explained the position of assistant to which Crookes was appointed and which he retained till 1854. Mr. John Spiller, afterwards prominently concerned in the manufacture of aniline dyes, entered the College of Chemistry at the same time as Crookes, and the two young students became close friends. Both were interested in photography, and together published various papers on the preservation of sensitised collodion plates and other cognate subjects. In 1854 Crookes went to Oxford as photographer to the Meteorological Department of the Radcliffe Observatory, and in 1855 he became Lecturer on Chemistry at the Chester Training College.

In 1856 he married Miss Ellen Humphrey, a native of Darlington, whose acquaintance he had made some years earlier as a school-friend of his cousin. They were married on April 10, 1856, at the parish church of St. Pancras, Middlesex, and began housekeeping at Brompton. They removed to 20, Mornington Road, N.W., and afterwards to 7, Kensington Park Gardens in 1880, and this was their home to the end of their lives. In 1859 he brought out the first number of the *Chemical News*, of which he remained sole editor until 1906.

About this time the employment of the prism in recognising and distinguishing volatile substances in flame was introduced by Bunsen and Kirchhoff, and the discovery of rubidium and cæsium in the water of the Dürkheim spring was announced by Bunsen in 1860. On applying the spectroscope to the seleniferous material from the vitriol works at Tilkerode (Harz), which had been given to him by Hofmann some years previously, and which he had used as the source of selenium in his work on the selenocyanides, Crookes observed a new green line which led him to the discovery of thallium. The first announcement of the existence of a new element appeared in the *Chemical News*, March 30, 1861. It was originally supposed to be related to sulphur, but the discoverer soon saw fit to alter his opinion, and the specimens exhibited in the International Exhibition, 1862, and to which a medal was awarded, were labelled "Thallium, *a new Metallic Element.*" The discovery was interesting from several points of view. The use of the spectroscope was novel and the properties of the new metal were strange, exhibiting as it does the appearance and approximately the density of lead, some of its salts resembling those of lead and mercury, whilst others are perfectly similar in solubility and crystalline relations to the salts of the alkali metals. The discovery of thallium at once secured for William Crookes a recognised position in the scientific world, and in 1863 he was elected F.R.S. Obviously the first task which lay before him was the investigation of the chemical and physical properties of the new element and its sources in nature. Among the most important of the physical constants to be determined was the atomic weight, and the remarks on the subject contained in Prof. F. W. Clarke's *Constants of Nature* (Smithsonian Institution, 1882) may properly be quoted here. "In 1873, Crookes, the discoverer of thallium, published his final determination of its atomic weight. His method was to effect the synthesis of thallium nitrate from weighed quantities of absolutely pure thallium. No precaution necessary to ensure purity of materials was neglected; the balances were constructed specially for the research; the weights were accurately tested and all their errors ascertained; weighings were made partly in air and partly *in vacuo*, but all were reduced to *absolute* standards, and unusually large quantities of thallium were employed in each experiment. . . . Suffice it to say that the research is a model which other chemists will do well to copy. . . . Hence, using the atomic weights and probable errors previously found for N and O, Tl = 203·715 + ·0365. If O = 16, Tl = 204·183. . . . Crookes himself, using 61·889 as the molecular weight of the group NO_3 gets the value Tl = 203·642; the lowest value in the series being 203·628 and the highest 203·666, an extreme variation of 0·038. This is extraordinary accuracy for so high an atomic weight, at least as far as Crookes' work is concerned."

This passage illustrates the spirit which animated Crookes' work throughout. Nothing short of the highest attainable accuracy ever satisfied him.

Ordinary research into chemical and physical phenomena was at this time interrupted to some extent owing to excursions during the years 1870 to 1873 into a field which up to that time had been regarded as outside the domain of pure science. But Crookes, with characteristic independence of spirit, plunged into an enquiry from which his contemporaries in pure science not unnaturally shrank. The result was that an article under the title "Experimental Investigation of a New Force" appeared from the pen of William Crookes, which created a feeling of consternation among his scientific friends. Mr. D. D. Home had been for some time holding séances at which his presumed powers as a "medium" had been manifested in the production of phenomena which could only be described as extraordinary. These have been so often described that it is only necessary here to mention the production of sounds like raps or crackling, the movement of heavy bodies such as furniture in the presence but out of reach of the medium, and the "levitation" or apparent floating in the air of the medium or other human beings. The experiments made by Crookes were stated by him to have been conducted in his own house, in the light, at times appointed by himself and in the presence of no person beside the medium, except private friends.

At one of the earlier séances the experiments were made in the presence of Dr. Huggins (afterwards Sir William Huggins,

President of the Royal Society), Serjeant Cox, proprietor and conductor of the *Law Times* and *Recorder of Portsmouth*, one of Crookes' brothers, and his chemical assistant. Later, in January, 1874, the same journal published "Notes of an Enquiry into the Phenomena called Spiritual during the years 1870-73," with the signature William Crookes.

It is perhaps not surprising that Crookes was publicly attacked in a violent manner, but he was able to show that many mis representations and misstatements were made which everyone must now perceive were wholly unjustifiable. The story of his experiences as told by him is supported by evidence which would be accepted as conclusive if these statements related to any scientific work or to any ordinary occurrence. Crookes himself never withdrew or altered his statements concerning the phenomena he had witnessed, and in his Presidential Address to the meeting of the British Association at Bristol, so late as 1898, he reiterated his conviction as to their reality. This conviction he retained to the end of his life. He was President of the Society for Psychical Research in 1897. His view, if he really had a settled opinion, as to the explanation of these strange phenomena cannot be given in his own words, but the view of Mr. Serjeant Cox on the theory of what he called psychic force is given very clearly at the end of these "Notes," and it appears probable from the prominence given to this exposition that it represents very nearly the opinion of Crookes himself. Perhaps the last few lines are sufficient to quote in this place, as probably views may have changed during the forty-five years since they were written. The passage is as follows: "The difference between the advocates of psychic force and the Spiritualists consists in this —that we contend that there is as yet insufficient proof of any other directing agent than the Intelligence of the Medium, and no proof whatever of the agency of the Spirits of the Dead; while the Spiritualists hold it as a faith, not demanding further proof, that the Spirits of the Dead are the sole agents in the production of all the phenomena. Thus the controversy resolves itself into a pure question of *fact*, only to be determined by a laborious and long-continued series of experiments and an extensive collection of psychological facts."

With regard to the phenomena recorded by Crookes there will be for a long time to come two opposite and irreconcilable opinions. On the one hand the sights and sounds exhibited in the presence of the medium are not more inexplicable than many of the tricks shown by professional conjurers in the production of which there is no claim as to the employment of "psychic force" or any supernatural agency. The late Mr. Bertram, a professional illusionist, used to show a trick with *solid* brass rings which were made to link together and unlink *immediately under the eyes of the spectator*, to whom it appeared that the metal of the one ring passed through the solid metal of the other. The essence of the illusion here has never been explained to the public, but it is known to the conjuring confraternity, and no one suspects them of any supernatural power.

On the other hand many people now believe that the agency of the "medium" provides a means of communicating with the spirits of the departed, and there can be no doubt that such persons have in many cases become unconscious victims of cruel fraud, rendered all the more easy by the general prevalence of a kind of hyper-sensitiveness following the agonies of the late war.

It is unnecessary to pursue the subject further, but as Crookes made no secret of his views, and his credibility in regard to all other questions, scientific or otherwise, has never been impugned, his biographer would not be justified in doing more or less than to place on record such statements as appear to represent fairly the position he had assumed, and certainly no biographer would be expected to pronounce any opinion other than that which he believes to have been entertained by the subject of his notice.

Returning to the chief events in Crookes' scientific career, the estimation of the atomic weight of thallium involved the use of a balance contained in a case exhausted as completely as possible of air. The use of the vacuum balance was attended by unexpected phenomena, which occupied his attention for many years afterwards and led to the discovery of the *radiometer* in 1875. This was described in a paper entitled "On Attraction and Repulsion resulting from Radiation," communicated to the Royal Society on March 20, 1875. A Royal Medal was awarded to Crookes at the Anniversary Meeting in the same year. The President in presenting the medal referred to the instrument in the form in which it has been since familiar, namely, with the four-armed fly mounted on a sharp point and having the vertical discs at the ends of the arms blackened on one side. He also remarked, "It is the mystery attending this phenomenon that gives it its great importance." Great interest was manifested by

Sir William Crookes' Physical Laboratory.

many experimenters in the phenomena observed, and there were many attempts at explanation. In the end, the hypothesis put forward by Dr. G. Johnstone-Stoney, according to which the repulsion is due to the movements of the molecules of the residual gas acting differentially on the two surfaces of the movable disc, was accepted.

In a footnote to one of his papers (*Proc. Roy. Soc.*, 1876, **25**, p. 308), Crookes drew attention to the properties of highly attenuated gas, and expressed the view that the phenomena indicate the existence of a fourth state of matter as far removed from the condition of gas as gas is from liquid.

In all the numerous experiments connected with this investigation, Crookes was assisted by Mr. C. H. Gimingham, whose mechanical dexterity and skill as a glass-blower were quite remarkable. Gimingham joined the Swan Electric Light Company in 1881, but unhappily died a few years later.

The phenomena exhibited by the electric discharge in rarefied gas had long been familiar, and had been studied by Plücker, by Hittorf, and other physicists. It was natural that in the examination of the properties of highly attenuated gas the phenomena exhibited by electric discharge through such media should receive Crookes' attention, and in the paper in which his first experiments in this direction were described (*Proc. Roy. Soc.*, 1879, **28**, p. 110), he was led to theoretical speculations on the ultra-gaseous state of matter. In this paper, the dark space which appears round the negative pole was the subject of experiment, and was found to enlarge as the exhaustion proceeds, whilst the phosphorescence excited on the glass walls of the tube diminishes and ultimately disappears. The dark region round the electrode has since been known as the *Crookes, or cathode dark space.* The rays from the cathode may be made to converge by the use of an aluminium cup, and the result is the production of a green phosphorescent spot on the glass; the rays, travelling in straight lines, cast a strong shadow from any object placed in their path. The rays when concentrated also develop great heat, which may rise to the melting point of platinum.

In 1880 the French Académie des Sciences awarded Crookes an extraordinary prize of 3,000 francs and a gold medal in recognition of his discoveries in molecular physics and radiant matter. In the years following 1880 he continued the active investigation of the phenomena exhibited by gases in a highly attenuated state, and among other properties he studied the heat conduction and viscosities of gases in this condition. He also examined the phosphorescence exhibited by many substances when exposed to the discharge from the negative pole in a highly exhausted tube; and in the Bakerian Lecture for 1883, speaking of this discharge as "radiant matter," he considered that the particles flying from the cathode were of the dimensions of molecules.

For some years he was occupied in tracing by the spectroscope the changes noted, and he was led to attempt the separation of some of the earths, notably yttria, into the components of which they were supposed to consist, by means of a very elaborate system of chemical fractionation. The result of all this work led him to speculations as to the characters of the elements and the existence of a class of bodies which he called "meta-elements." These meta-elements he regarded as composed of atoms "almost infinitely more like each other than they are to the atoms of any other approximating element. It does not necessarily follow that the atoms shall all be absolutely alike among themselves. The atomic weight which we ascribe to yttrium therefore merely represents a mean value around which the actual weights of the individual atoms of the 'element' range within certain limits. But if my conjecture is tenable, could we separate atom from atom we should find them varying within narrow limits on each side of the mean." This view was put forward in the Presidential Address to the Chemical Society, 1888, the whole of which even now, after more than thirty years, would repay perusal. The possibility of the evolution of the elements from a primal elementary protyl or *urstoff* is a proposition which has been discussed from the most ancient times, but the chemist had little positive information as to the inter-relations among the recognised elements before the conception of the periodic law. Taking an idea from Prof. Emerson Reynolds for the diagrammatic display of the periodic relation of properties to atomic weight, Crookes produced a figure of eight curve, on which the symbols of the elements are placed at intervals so that the members of natural families fall into position vertically over one another.

On this curve the meta-elements would be ranged in groups or clusters close together. Radioactivity had not been discovered when this address was composed, but something approaching

SIR WILLIAM CROOKES' CHEMICAL LABORATORY.

Crookes' idea has been realised in more recent times by the discovery of isotopes among the products of the disintegration of radioactive elements. For his researches on the behaviour of substances under the influence of the electric discharge in a high vacuum, with special reference to their spectroscopic behaviour, the Davy Medal was awarded to Crookes by the Royal Society in 1889.

The discovery of argon by Rayleigh and Ramsay in 1894, and of helium by Ramsay in 1895, opened a new field, and Crookes being recognised as the most experienced observer of spectra and the highest English authority in this direction, the new gases were at once submitted to him, and the identity of terrestrial with solar helium was established finally by his examination.

Soon after this time the brilliant researches of J. J. Thomson threw an entirely new light on all the difficult spectroscopic questions which had been so patiently and so successfully examined by Crookes. It became clear that the particles projected from the cathode were much smaller than any known atoms or molecules of ordinary matter, and were, in fact, the minute bodies called electrons. On this point, Crookes, in a paper on the "Stratifications of Hydrogen" (*Proc. Roy. Soc.*, 1902, **69**, p. 411), expressed himself in the following terms: "In twenty-five years one's theories may change, although the facts on which they are based remain immovable. What I then called 'Radiant Matter' now passes as 'Electrons,' a term coined by Dr. Johnstone-Stoney to represent the separate units of electricity which is as atomic as matter. What was puzzling and unexplained on the 'Radiant Matter' theory is now precise and luminous on the 'Electron' theory." And by application of this theory the stratifications of hydrogen were explained.

The discovery of radioactivity by Henri Becquerel and of radium by Madame Curie toward the end of the century naturally attracted Crookes' interest and attention. On examination of uranium salts, some specimens were found to be much more radioactive than others. It was soon found that the radioactive constituent is precipitated from a solution by ammonia, and on adding excess of the reagent, a small, insoluble, light brown precipitate is left which exhibits strong radioactive properties, whilst the uranium salt remaining is almost inactive. A year later the uranium had regained its activity. To the active substance the name uranium-*X* was given. Whatever be its nature, it is evident that uranium owes its usual activity to the presence of this substance which is generated from it, and the separation of which depends on the readiness with which it attaches itself to precipitates, especially ferric hydroxide, when iron is present as an impurity in the uranium compound employed as material. The year following, Crookes, continuing his observations on the emanations of radium, discovered the effects produced by the α-rays on a surface of hexagonal blende (zinc sulphide), and invented the little instrument, which he called the spinthariscope, by which the number of scintillations can be counted, each spark being produced by one α-particle.

The researches on the spectra and other characters of the rare earths occupied more than twenty years, and one result was a very extensive study of scandium and its salts, which places it in the position of being now better known than almost any other of these difficult elements from which scandium is separated by its low atomic weight.

A very valuable investigation undertaken in connection with the Glass Workers' Cataract Committee of the Royal Society was begun in 1909. The main object was to prepare a glass which would cut off those rays from highly heated molten glass which damage the eyes of workpeople. This involved a study of the effects of the addition of a large number of metallic oxides to a colourless glass, specially prepared for the purpose by Mr. H. Powell, of the Whitefriars Glass Works. The problem was to prepare a glass which would cut off as much as possible of the heat radiation, and at the same time be opaque to the ultra-violet rays, whilst the colour would be scarcely noticeable when used in spectacles. In the result, a series of eighteen different recipes were provided which meet, more or less fully, the three requirements contemplated. These have been found to be, in practice, very beneficial to the workers.

We may now recall some of the subjects which at various times were studied by this untiring worker, outside the course of research which may be supposed to have represented his predilection.

It should not be forgotten that when quite a young man he was appointed by the Government to report on disinfectants and their application to the arrest of the cattle plague in 1866, and that he was responsible for the recognition of the antiseptic value of phenol or carbolic acid.

The diamond has been ever a subject of interest, not merely to the jeweller, but to the mineralogist and chemist, owing to the mystery in which, up to recent times, its origin and formation were involved. In a visit to Kimberley in 1896, Crookes spent nearly a month in the mines studying the question of the origin of the mineral, and again in 1905, on the occasion of the visit of the British Association, he pursued the same inquiry. In 1893 the late Prof. Moissan demonstrated the production of diamonds by crystallisation of carbon from molten iron under pressure, and Crookes showed that the residue of cordite exploded in a closed steel cylinder contains crystalline particles possessing the form of the diamond. He published an interesting little book on diamonds in 1909.

Another subject in connection with which he did good service was the importance of producing and applying to the land much larger quantities of nitrogen in the form of nitrate in order to increase the supply of wheat. "The Wheat Problem" was one theme of his Presidential Address to the British Association at the meeting at Bristol in 1898, and his views on the subject were embodied in a volume published a year later, in which he was able to reply to the various critics who in the meantime had questioned some of his conclusions. Though, doubtless, some of his most startling statements admit of modification, the problem still remains a topic of supreme interest to the agriculturist and to the world at large.

It would be almost impossible to enumerate all the various directions in which Sir William Crookes (he received the honour of knighthood in 1897) occupied himself in connection with problems of public interest or as expert adviser to the Government, but, in passing, may be mentioned his work on the disposal of town sewage, his reports on the composition and quality of daily samples of the water supplied to London from 1880 to 1906, and his services as Consulting Expert on the Ordnance Board from 1907 onwards during the period of the war. Nor should it be forgotten that the office of President is in many learned societies no sinecure. In presiding over the Chemical Society (1887-9), the Institution of Electrical Engineers (1890-4), the British Association (1898), the Society of Chemical Industry (1913), and, finally, the Royal Society (1913-15), Sir William paid close attention to all the multifarious details of the business of each society. He also served as Honorary Secretary to the

Royal Institution from 1900 to 1913, and as Foreign Secretary to the Royal Society from 1908 to 1912. Every man of science among his contemporaries will be ready to affirm, therefore, that the numerous honours which were showered on Crookes by the most distinguished academies and universities in the world were well earned and very fittingly conferred. He received from the Royal Society the Royal, the Davy, and the Copley Medals, and from the Royal Society of Arts the Albert Medal, and finally, in 1910, the Order of Merit was conferred on him by the King.

Crookes' whole scientific career is interesting, apart from the value of his discoveries, as illustrating the fact that to a man of genius the character of his early education has but little influence on his achievements. As mentioned already, he left school at the early age of fifteen, and at once specialised in a single branch of science under a teacher eminent in his own line, but from whom the young student seems to have derived little but the advantage of example, for the subjects to which Hofmann devoted his energies appear to have had but little attraction for Crookes. Unlike W. H. Perkin, who also entered the college at about the age of fifteen with equally imperfect general education, he never seems to have been attracted by organic chemistry, and to the end of his life remained practically ignorant of this branch of science. The genius displayed by Crookes was, however, accompanied by unusual independence of character, which was displayed in a variety of ways, not only in the course taken by his own researches, but by his attitude toward the statements and pretensions of others. Nothing seemed too improbable to escape his attention, and of this the time and trouble he was tempted to expend on the pretended transmutation of silver into gold some twenty years ago is a sufficient illustration. The same liberality of spirit made him very tolerant, and perhaps not always sufficiently critical, in regard to articles in his paper, the *Chemical News*.

Crookes' industry was extraordinary. His first paper, already mentioned, was published in 1851 and the last in 1918, a period of sixty-seven years devoted almost entirely to research. But he published also several bulky volumes of technical character concerning which perhaps the jingling rhymes by Frederick Field in the *Chemical Review* by "a B" of long ago represent a not unfair judgment:

> "Thallium, we hail thee, and we owe to Crookes
> More for thy happy birth than for his books."

They undoubtedly represented a considerable amount of labour, probably performed to some extent by assistants.

Crookes was no linguist, and possibly owing to the fact that his researches were never contributed to any German journal his work was very little noticed in Germany. It was also evident from the reception given to his address delivered at the Chemical Congress in Berlin in 1903 that the audience was unfamiliar with his person and indifferent to his subject. Nor was he much interested in literature, and in evidence of this he once confided to a lady with whom he was talking on the subject that his recreative reading generally was found in detective stories. This is not to be surprised at, for the problems before the detective are of a character similar to those which the man of science endeavours to elucidate. Crookes was kindly in disposition, helpful to young men and hospitable toward his friends and acquaintances. When he came to live in Kensington Park Gardens he was always at home to callers on Sunday evenings during the winter, and it was on one of these occasions that the writer had an opportunity toward the end of the century of learning from his host that his convictions were unchanged regarding the reality of the phenomena of levitation, etc., he had recorded thirty years previously. He was, in fact, at that time still actively engaged, as President of the Society for Psychical Research, in following up the subject.

Spare in figure throughout his life, he became toward the end extraordinarily thin, though his capacity for work when over eighty years of age was illustrated in a remarkable way by his active discharge of the duties of President of the Royal Society, and at the same time, in 1913, those of President of the Society of Chemical Industry. Few can have had opportunities of witnessing any disturbance of his temper, which was singularly equable, but it must also be said of him that he took life seriously, for he was one of those who are "never seen to laugh and rarely seen to smile." This is not inconsistent with an habitually cheerful demeanour, to which continuous domestic happiness doubtless contributed in no small degree. The dedication of his little book on Diamonds is an expression of acknowledgment: "To my Wife, my companion and friend of fifty-four years. To her judgment and advice I owe more than I can ever repay." Lady Crookes died in 1915.

Whatever position in the hierarchy of science is ultimately

assigned to Crookes, there can be no doubt that his work on the electric discharge in gases will determine the question.

The discovery of thallium was a brilliant result of the application of a new experimental method, but it would certainly have been made within a few years by one of the many chemists who resorted to the use of the spectroscope. Crookes' laborious experiments in the fractionation of the rare earths led to no discovery comparable with the separation of neodymium from praseodymium by Auer v. Welsbach in 1885. Nevertheless, it was always instructive to hear of Crookes' views on any subject, whether it was the origin and relations of the elements or the ultra-gaseous state of matter. And so the judgment of the contemporary generation of chemists must certainly place him in the front rank of scientific pioneers.

Yours sincerely

William Ramsay

[Face p. 273]

CHAPTER XXI

RAMSAY

THE name and scientific achievements of SIR WILLIAM RAMSAY are still fresh in the memories of his scientific contemporaries and the public. He died in the early morning of July 23, 1916, in the midst of the great struggle in which his country was involved, and at a time when his great ingenuity and activity of mind might have been of service. He had already made strong representations to the Government on the absurdity of allowing fats and cotton, both materials from which explosives were made, to pass into the hands of the enemy through some of the nations then assuming a position of neutrality. After some nine months of suffering he was, however, removed from further pain and further activity,

> "Like a summer-dried fountain, when our need was the sorest."

William Ramsay came of a stock from which mental qualities might be expected such as would lead to diligent pursuit of some branch of science, probably chemistry. His forefathers for many generations had been dyers on the paternal side, while those on the mother's side had been physicians. His father, William Ramsay, born in 1811, had married Catherine Robertson after waiting many years, both being about the age of forty, and their only child, William, was born at Queen's Crescent, Glasgow, on October 2, 1852.

From notes supplied by friends of the family we learn that young Ramsay had a very happy childhood. From the fact of his being an only child much of his time was spent in the society of his parents, and hence he acquired the habit of speaking in the style of his elders, and probably his thoughts were to some extent influenced by what he heard from them. He did not care greatly for the games on which boys generally spend so much of their time. He was, however, fond of boats and building with bricks, according generally to plans of his own. He had a strong love of animals and always had a dog as pet and companion. From his earliest days he showed considerable aptitude for music, and he had instruction in playing the piano not only at the time when he was attending a preparatory school, but in later years from Dr. A. L. Peace, organist of Glasgow cathedral. He also developed remarkable powers as a linguist, which he cultivated while still a young boy by the practice of reading a French or German Bible in church while the sermon was going on. His mother was a strict Calvinist, and her boy was therefore brought up under influences which gradually underwent modification as he advanced in life, but which had one great advantage in making him familiar with the text of both Old and New Testaments.

His grandfather's family consisted of one daughter, Eliza, born in 1810, and three sons, William, born in 1811, Andrew Crombie (afterwards Sir Andrew, the famous geologist), born in 1814, and John, who became a sugar-planter in Demerara, born in 1816. On this side the cousins of young William were the children of the geologist, and lived in London. On the other side the only sister of his mother had married the Rev. Dr. Jolly, and their only son became a minister in Shetland. It was in visits to these relations in the early years, accompanied by his mother, that young Ramsay gained an extensive familiarity with his native land. In Shetland he learned to swim and to manage a boat in all weathers. He also enjoyed visits to the family of the Turnbulls, one of whose ancestors had been a partner with his grandfather, and was the discoverer of "Turnbull's Blue." They had been manufacturers of pyroligneous acid and other products from the distillation of wood.

In due time William Ramsay entered the Glasgow Academy, and of his life at this period the recollections of his old schoolfellow and lifelong friend, Mr. H. B. Fyfe (since deceased) supply interesting glimpses. He says: "I met William Ramsay for the first time in August, 1863, when we both joined the Third Latin Class in the Glasgow Academy. . . . He continued in the Glasgow Academy till May, 1866, taking the Third Latin in '63-64, the Fourth Latin in '64-65 and the Fifth Latin in '65-66. So far as I remember he did not take any part in the class games, and I do not remember that he took any prizes. I think this was accounted for by the fact that he was nearly two years junior to the average age of the class. . . . He was, all the same, about the average height. In November, 1866, he went to the University and took the usual Arts Degree curriculum. This would be, I think, Latin and Greek for two years, that is till May, 1868, Logic and Mathematics in 1868-69, and Natural Philosophy and Moral Philosophy in 1869-70. I do not think he ever took Chemistry in the University.

"Our friendship began shortly after we went to College. . . . At that time he knew nothing of chemistry theoretically, but he had for some time been working at home at various experiments, as we called them. He worked in his bedroom and there were a great many bottles always about containing acids, salts, mercury and so on. When we began to meet I found he was quite familiar with all the ways of getting the material and apparatus for working in chemistry. We used to meet at my house in the afternoons and do what practical work we could, making oxygen and hydrogen and various simple compounds such as oxalic acid from sugar. We also worked a good deal with glass. . . . We made nearly all the apparatus we used except flasks, retorts, and beakers. . . .

"He did not begin to study chemistry systematically till about October, 1869, when he went to Mr. Tatlock's laboratory in the afternoons, after his college classes for the day were over, and began to work under him. . . . In addition to his College classes and chemistry he took lessons in music under Dr. Peace and also in French and German, the German being under Dr. Schlomka.

"It was in 1869 that he first went to stay with his cousin Mr. Jolly at Walls. . . . In the following winter '69-70 my eyesight failed and for nearly a year I was unable to read. During that time he was exceedingly kind to me in coming to read to me, and also in arranging for walks in the country. He must by that time have acquired considerable familiarity with languages, for he read a good deal of Béranger and other French poets. He was also at the same time trying to learn Gaelic, and used to amuse us a good deal by going down to the kitchen to test his progress and to ask questions from a Highland cook they had. . . .

"Of course I met Ramsay's father and mother frequently and stayed with them often. I think Ramsay took his scientific bias from his father, who was greatly interested in every scientific subject, and glad to talk about them. . . . I think he also took his sociable disposition and broad outlook on life a good deal from his father.

"His mother was a very notable woman. She had very strong views on all matters connected with religion and the Church, and I think she looked with considerable suspicion on scientific enquiries. She did not oppose or argue about such enquiries, but she certainly did not encourage talk about such subjects as Darwinism, and was immediately up in arms if anything was said that seemed to question orthodox religious views. I think it must have been owing to her that so much of Ramsay's education was devoted to such subjects as Latin and Greek and Philosophy, in which he never took any particular interest. She had splendidly calm nerves and a clear logical head, both of which qualities she passed on to her son. . . .

"When he came back from Germany he was splendidly equipped both mentally and physically for a successful career. He had perfect health. He could walk forty miles in a day without any difficulty. He was a very strong and graceful swimmer and could dive further than any amateur I have seen. When we were in Paris in 1876 the four of us used to go to one of the baths in the Seine every forenoon, and after the first time, when Ramsay was ready to dive the bathman would pass round the word that the Englishman was going to dive, and everyone in the establishment, including the washerwoman outside, would crowd in and take up positions to watch him. He dived the whole length of the bath and sometimes turned there under water, and came back part of the length. . . . With these physical qualities he had also great courage and tenacity. He never seemed to be upset by anything. . . . I always found him exceedingly kind-hearted and considerate. I do not think anyone who needed friendliness or consideration would ever have been passed by by him, and for my own part, especially at the time when my eyesight failed me, he was kindness itself."

In Tatlock's laboratory a year was devoted exclusively to analytical work, and in the following year Ramsay attended the lectures in chemistry by Prof. Thomas Anderson and his successor, John Ferguson. He also, among other subjects, attended lectures and laboratory instruction in physics under Sir William Thomson (Lord Kelvin), of whom in an essay written in later years he speaks in terms of admiration and gratitude for his own indebtedness to the teaching and example of his great fellow-countryman. The project was formed about this time of studying chemistry in Germany, and at first it was intended that he should go to Bunsen at Heidelberg. The outbreak of war between France and Germany caused this project to be deferred. He did, however, find his way to Heidelberg in the autumn of 1870, and appears to have worked for a short time in the laboratory there, but in the spring of 1871 he went to Tübingen and entered the laboratory of Professor Fittig, where in the following year he secured the degree of Ph.D. He enjoyed the society of a number of English, Scotch and American students, among whom was Ira Remsen, afterwards Professor Remsen, of Baltimore, who at that time was acting as an assistant in the laboratory.

Returning to England, Ramsay was appointed assistant to the Professor (Bischof) of Technical Chemistry in Anderson's College, Glasgow.

In the summer of 1873 Ramsay joined his uncle Andrew, who, with his sister Eliza and eldest daughter, were visiting the Rhine valley with geological investigation in view. With this eldest cousin of his, Ella Ramsay, he kept up a correspondence throughout his life, and from his letters, which have been preserved, much information concerning all the principal events of his career is derived.

In 1874 Ferguson succeeded Anderson as Professor of Chemistry in the University, and Ramsay was appointed to the post of Tutorial Assistant. This he held for about six years, and after offering himself for several collegiate vacancies, he was ultimately successful in his candidature for the Professorship of Chemistry in the then newly-established college at Bristol. This after Ramsay's time developed into the University of Bristol.

He entered on the duties of this appointment early in 1880. Those were days when the country was slowly waking up to the importance of providing teaching of high quality within reach of those who for various reasons were unable to take advantage of the benefits to be enjoyed at the older residential Universities. The conditions under which the new professorships were called into existence were not always encouraging and attractive to men of ability. The community in many cases was indifferent, the funds inadequate, and the ideas of the governing bodies on such subjects as teaching and research were narrow; the work imposed on the teachers was often too heavy, so that there was scanty time for self-improvement and the pursuit of new knowledge. In the case of Bristol the stipend offered to the Professor of Chemistry was £300 per annum, with a share of the fees paid by students, a minimum of £400 being guaranteed by the Council. But the Professor was required to give courses of lectures in College both to day and evening classes, also to give courses of instruction in neighbouring towns in connection with local industries. Thus by an arrangement with the Cloth-Workers Company he had to lecture in Trowbridge on the art of dyeing. The first Principal of the College was Alfred Marshall, afterwards Professor of Political Economy at Cambridge, but he resigned in 1881 and Ramsay was appointed in his place. This involved an additional burden of duties, but Ramsay was never afraid of work and the laboratory was kept busy with research in which he was joined by his successive assistants. The chief of these was Dr. Sydney Young, who was his successor in the Chair and afterwards became Professor in the University of Dublin. Between them a series of papers was produced on the thermal properties of solids and liquids and on the relation of evaporation to dissociation, which extended over and beyond the five years which elapsed before Ramsay left Bristol.

In 1884 the British Association met in Montreal, and Ramsay, together with several other Principals of the new University Colleges, was among the party on board the *Peruvian* s.s. on the return journey. The financial difficulties in which all these colleges were involved was naturally a subject of discussion among them.

The Welsh Colleges had recently obtained grants of money from Parliament, and the idea arose that the English Colleges, which were all more or less hampered by want of money, should receive similar assistance. Ramsay and Principal Hicks, of Sheffield, were appointed secretaries of the Committee which was soon afterwards formed to take steps in this direction, and Ramsay particularly took a very active part in all the business, which happily resulted some years later in the assignment of a grant from the estimates in aid of university education in the provinces.

In 1887 Ramsay was chosen to succeed Williamson in the Chair of Chemistry at University College, London. Immediately afterwards he was placed among the fifteen selected for the F.R.S. In this position at University College he remained till his

Sir W. Ramsay Lecturing at University College, London.

retirement in 1912, and here all his remarkable and epoch-making discoveries were made. As a teacher Ramsay held very strong views, first as to the place and manner of examinations in any system of education in science, and secondly as to the importance of introducing students to the methods of research at the earliest possible moment. On these points there will always be considerable difference of opinion among those best qualified to judge, that is, among teachers of experience. But whatever his views he was certainly very successful in inspiring his senior students with his own ardour in the pursuit of new knowledge, and if the rank and file did not distinguish themselves under examination it was not owing to neglect by the professor. He was, in fact, almost too ready to yield to the demand for fresh classes to suit the wishes of the College authorities. Ramsay had derived so much advantage from his time of study under Professor Fittig, at Tübingen, that he was led to attach somewhat exaggerated importance to the principle of letting teachers examine their own students with very little external censorship. This is not the place to discuss a question of this kind, but those who are interested will find in the Life of Sir William Ramsay [1] a chapter in which his views are set forth in some detail with the aid of extracts from his own letters and essays.

Before Ramsay had been long at University College an occasion arose which led to his great discoveries in connection with the gases of the atmosphere. During the intervening five years he had been carrying on work of physico-chemical character which resulted in a number of published papers, also in the business connected with efforts then being made to change the character of the London University.

In 1892 Lord Rayleigh, in the course of a long series of experiments on the densities of the principal gases, discovered the fact that nitrogen gas derived from the air is slightly heavier than nitrogen prepared by a chemical process from ammonia or other compound containing that element. In the two following years an investigation into the possible causes of this difference led to the conclusion that it must be due to the presence in the atmosphere of a small quantity of a gas heavier than nitrogen

---

[1] *Sir William Ramsay, Memorials of His Life and Work*, by Sir William A. Tilden (Macmillan. 1918).

which remains with the nitrogen in all those operations, by which the moisture, the carbon dioxide and the oxygen are removed and the nitrogen left. Lord Rayleigh's attention had been drawn to the old experiments of Cavendish described in 1785, and of which an account is given on p. 44 , in which with characteristic caution Cavendish had stated that if there was any portion of the air different from that which entered into the composition of nitre it was certainly "not more than $\frac{1}{120}$th part of the whole." Lord Rayleigh made use of Cavendish's method; that is, he subjected a limited volume of air to an electric discharge, while supplying oxygen in the presence of an alkaline solution which served to absorb the oxides of nitrogen as fast as they were formed, leaving the unabsorbed gas mixed with the excess of oxygen employed. Lord Rayleigh had appealed to the chemical world for suggestions, and Ramsay was the only chemist to respond by practical experiment. He made use of the property possessed by red-hot magnesium of combining with nitrogen, forming a solid compound, a nitride. This method, arranged so as to be applied in the form of a continuous process, was more rapid than the Cavendish method, and the result was that by the end of July, 1894, the gas afterwards named argon was isolated and its chief properties recognised. Argon is a colourless gas distinguished from all others then known by its density, represented approximately by the number 20 (H = 1), and by its inability to enter into any form of chemical combination. Ramsay soon began to look among the rarer minerals for evidence of the existence of compounds of argon, and his attention was drawn to the fact that the natural uranites on being dissolved in acid frequently give off a gas which had always been supposed to be nitrogen. On examination of this gas Ramsay found that it contained not only argon but the gas *helium*, an element supposed to exist in the sun and the characteristic spectrum of which had been observed by Lockyer many years before. Helium, like argon, is chemically inert.

The study of the position indicated in the periodic table of the elements for these two remarkable substances led Ramsay to look for others of the same family, and in 1898 the search was rewarded by the discovery of three others, *viz.*, neon, krypton and xenon. These five gases form a continuous series which stands between hydrogen and the halogens on one side, and lithium and the alkali metals on the other, as follows:

Hydrogen 1	Helium 4	Lithium 7	Glucinum 9
Fluorine 19	Neon 20	Sodium 23	Magnesium 24
Chlorine 35·5	Argon 40	Potassium 39	Calcium 40
Bromine 80	Krypton 82	Rubidium 85	Strontium 87
Iodine 127	Xenon 128	Cæsium 133	Barium 137

The atomic weights are given here in round numbers.

Since 1895, when helium was discovered as a constituent of many minerals, it has been found present in varying quantity in practically all natural gases and spring waters. It is also present in atmospheric air in minute proportion. The density of this gas is only twice that of hydrogen, and being, unlike hydrogen, incombustible, it would be very valuable for inflating balloons and airships if obtainable in sufficient quantity at a cost which would not be prohibitive. In 1915 an investigation was started by Professor McLennan of the University of Toronto, at the instigation of the British Admiralty, with the object of ascertaining the proportion of helium present in various samples of natural gas and of devising a method by which the precious gas could be isolated in large quantity. Natural gases from Ontario and Alberta, Canada, were found to be richest in helium, and these sources were calculated to yield from ten to twelve million cubic feet per year. The process of separation consists essentially in freezing out the hydrocarbons, the carbon dioxide and the nitrogen which form the bulk of the natural gas as it issues from the ground, leaving the helium which was present to the extent of 0·33 per cent., uncondensed and nearly free from other gases. The machine by which this result was accomplished has been described fully with diagrams by Professor McLennan, in a lecture delivered to the Chemical Society on June 17, 1920 (*J. Chem. Soc.*, **117**, p. 923, July, 1920).

Ramsay unhappily did not live to see this wonderful outcome of his discovery, but he did some things in connection with helium which remain unexcelled in interest from the philosophical and scientific point of view.

In 1902 the isolation of radium compounds by Madame Curie

opened a new field for speculation and research. The "emanation" always escaping from radium compounds was found to be an inert gas, like helium and argon, not permanent like those two gases, but undergoing decay from the moment of its liberation. With the co-operation of Dr. Frederick Soddy, Ramsay found on examination with the spectroscope that helium was a product of its decay, and in further experiments the volume of helium set free from a known volume of emanation was measured approximately. The density of the emanation was also determined by Ramsay in a truly remarkable series of experiments, and its molecular weight therefore settled. In view of its gaseous character and density Ramsay gave it the name *niton* and assigned to it a place in the series of argon:

Helium	Neon	Argon	Krypton	Xenon	?	Niton
4	20	40	83	130	176	222

Ramsay's views as to the possibility of radio-active emanations being capable of bringing about the disintegration of the common elements led him to make experiments, from the result of which he believed that he had succeeded in breaking down copper into an alkali metal, and lead into carbon. The difficulties in the way of establishing such results are very great, but in a different form of experiment Rutherford seems to have been more successful.

Ramsay was fond of travel, and several occasions arose, especially after his great fame was established, when business led him to undertake excursions to a greater distance than would come within the range of a summer holiday. An account of many of these journeys is to be found in the *Life* already referred to. Several of these, however, deserve mention even in this necessarily limited survey. The first was the occasion of the visit of the British Association to Montreal in 1884. After the conclusion of the meeting Ramsay and his wife made up a party, which included a brother of Mrs. Ramsay about to settle in Montana, for the purpose of visiting the Yellowstone Park.

In 1895, in company with Professor W. P. Ker of University College, Ramsay took a much needed holiday in Iceland, where the opportunity was not lost of collecting water from the hot springs at Reykjavik. The samples were taken home, but were found to contain no helium, although a notable quantity of argon.

In 1897, Ramsay being President of the Chemical Section of the British Association at Toronto, advantage was taken of the facilities granted to Members of the Association to proceed further westward and so to pay a second visit to Mr. Buchanan, who was now established on a ranch in Montana.

In 1900 Ramsay was requested by the Indian Government to go out to India and to advise on the best way to utilise the splendid gift offered by Mr. Jamsetjee N. Tata. The object in view was to set up an educational institution in India providing facilities for higher scientific work, beyond the range of the then existing Indian universities. One of the first questions to be settled was the locus of the new institution. Poonah and Bangalore were both possible sites and in the end the latter place was chosen. Portions of two letters, the second written on the way home, give an idea, though very imperfect, of the long journey and immense amount of work done between November, 1900, when Sir William and Lady Ramsay left London, and the end of February, 1901, when they returned home:

"BOMBAY,
"*December* 11*th*.

"We are in India! The exclamation mark is to express the absolute petrifaction and astonishment we both have at everything we see and smell. It is simply overwhelming. People of all colours from nigger black to pure white, dressed in all sorts of clothes from none at all, save smiles and a nose ring, to the most elaborate turbans, white overalls, and jewels of rare lustre, through all shades of white, yellow, orange, greens, blues, lilacs, and purples. No browns or sad colours. It is a spectacle that would make a colour-blind man curse his misfortune. The oddest groups; a woman with a naked child straddling her sideways, a scuddy of the male persuasion holding her hand, talking to an old gentleman, with spectacles and a white turban with a scarlet thread twisted in it, and bare legs, sitting on his hams, writing a letter for the lady. Two chaps with 'nozzings' on except a waist-cloth, and even exceptionally naked heads caused by an early and persistent use of the razor;—that's the kind of group that meets you everywhere. Old, old women, wrinkled and white-haired, with the remains of good looks about them, with the usual scanty garments, a saree or gauzy shawl, and a short petticoat. Fat, prosperous-looking, oily Hindoos, with white turbans and white blouses and trousers; Parsees with black bishops' mitres and European dress; in fact, one might go on for a week describing, and not exhaust one-tenth of the oddities one sees every moment. The crowds are prodigious, they swarm and struggle for space to stand or sit on. I am getting fluent in 'ghar ko jaloo,' etc. The most familiar part of Hindostani is the imperative mood, and indeed many Anglo-Indians never get beyond it. We have also a Goanese sprite, who goes and comes at our bidding. He is supposed to know English and readily understands what one wants, but his replies are often enigmas. However, with 'nods and becks and wreathèd smiles' we get along pretty well. You will wonder where education comes in here. Well, I have had long talks with Mr. Justice Candy, the Vice-Chancellor of the Bombay University, and with Mr. Giles, the Director of Public Instruction. I have seen a reasonably good technical laboratory run by —— as a private spec. He evidently has a lot of go about him. I have seen the Wilson College, where a sort of junior arts and science course is given. They too are dependent on the Bombay University, an examining body, with affiliated colleges after the fashion of the London University in 1859, when they were founded. I had to give a short account of what I intended to do. That is to go to Poonah on Sunday afternoon; spend Monday there; start on Monday night for Bangalore and arrive in thirty-six hours; thence to Madras, where we spend some days; thence to Calcutta by sea (Xmas on the ocean), and stay over the New Year at Calcutta. Then up the Ganges valley to Benares, etc., as far as Lahore, I think, and then back to Bombay by Allahabad. Padshah, the secretary, will not come with us. I want to see things with my own eyes, and not through his spectacles; he is theosophist, vegetarian, altruistic, and an admirable Crichton after the Indian model. We are going to see the Towers of Silence on Thursday. On Saturday we are to see the caves of Elephanta and also Jain temples. But the streets are the real attraction. We spent a couple of hours this morning wandering through the markets."

"P. & O. Steam Navigation Co.,
"SS. *Caledonia*,
"19. ii. 01.

"The site I have fixed on is Bangalore. The climate is excellent, neither too cold nor too warm. It is 4,500 feet up and there is a sort of fresh feeling like that on the top of a hill. There is a geological station there, an agricultural station, a college such as exists in India. They offer a splendid site 300 acres in extent, in the best part of the town, which is very open. They have £1,200 a year to play with, a sum which has fallen to the Mysore state as the result of a disputed legacy and which they wish to spend for the good of the state. The revenues of Mysore show a huge surplus which can't be annexed by the Maharaja, for he has a private allowance of £150,000. Also the Canvery Falls, 40 miles from Bangalore, are being connected with the Kolar goldfields 45 miles in the other direction and the leads pass within 7 miles of Bangalore. They are going to begin with 4,000 H.P., but measurements show that for eight months of the year 100,000 could be got. Lastly there are endless deposits of iron ore, manganese, magnesia, etc., near, all of which might be exploited. . . .

"The Committee have asked me to nominate the first members of staff, and I have suggested names."

In the year 1904 the Nobel Prize for Chemistry was awarded to Ramsay, while the prize for physics went naturally to Lord Rayleigh. They both went to Stockholm to receive the prizes from the hand of the King. In a letter to his friend Mr. Fyfe, Ramsay gave an account of their doings, from which the following passage is sufficient :

"We had a most gorgeous time for nearly a week, dining with all the celebrities, including old King Oscar. The old gentleman was very kindly and took Lord R. and me into his private room and showed us all his curiosities, the portraits of his sons when they were children and his reliques of Gustavus Adolphus and of Charles XII. The Crown Prince told Mag that it was a difficult job to be a king, thereby confirming the Swan of Avon. He said that whatever one supposed that a Norwegian would do, he invariably did the opposite. Indeed, there was nearly a bloodless revolution while we were there; the Prime Minister of Norway was there, and I believe the dilemma was only postponed."

In 1907 a visit was paid to Finland with the object of making the acquaintance of the family of Ramsays, who, perhaps originally Scots settled in Sweden, had for many generations lived at Helsingfors.

In 1904 Ramsay became president of the Society of Chemical Industry, and on this occasion the annual meeting was held in New York. This involved an address, and after the meeting the President and many of the members made a tour of visits to Philadelphia, Washington and other places, including St. Louis, where they visited the Exhibition and Ramsay gave a lecture on "Present Problems of Inorganic Chemistry" which was afterwards printed by the Smithsonian Institution. In 1908 and 1909 he also held office as President of the Chemical Society, and in 1909 he was chosen as President of the International Congress of Applied Chemistry, the meeting of which was held in London. At the previous meeting in Rome in 1906 there had been much discussion among the English delegates as to the choice of President, but the question was ultimately settled by appointing Sir Henry Roscoe, who was then an old man, Honorary President. Ramsay had very special qualifications for the office in consequence of his extensive personal acquaintance with chemists from all countries, his eminent services to science and his remarkable readiness in using foreign languages. This gift has already been hinted at, but it deserves to be remembered, for Ramsay himself was fully aware of the unusual quickness with which he was able to acquire a sufficient vocabulary to carry on communication in any new tongue. This he turned to account in the few days' journey to Iceland, in his visits to Sweden and Finland, and on the voyage out to India when he worked at Urdu.

At the meeting of the Congress which was held in the Albert Hall, with the Prince of Wales (King George V.) in the chair, Ramsay addressed words of friendly greeting to the foreign delegates in French, German and Italian successively.

In 1911 he was President of the British Association at Portsmouth.

After it was all over he went for a trip to Rio de Janeiro and back for a change and rest.

At the end of the summer session in 1912 he retired from the professorship at University College, having occupied the chair for twenty-five years.

He purchased a house and garden at Hazlemere, between two and three miles from High Wycombe, and having arranged a laboratory there he hoped to be able to continue his chemical researches.

In July, 1914, the British Association having gone to Australia, the Ramsays, with their son and daughter, contented themselves with accepting the invitation of the Association Française to the meeting at Havre. War was not expected on July 19th, when they made their plans, and even down to August 2nd while they were in Havre they expected to be able to stay at some place on the coast. Of course in a few hours they found that they must take the first opportunity of returning to England. Immediately after reaching home Ramsay threw himself enthusiastically into work connected with the national defence and offence, and he became a very active member of the Royal Society Committee. But the malignant disease which ended his life was already at work, and after several surgical attempts at cure, and months of pain, death brought relief. He lies in the churchyard at Hazlemere.

Together with his activity of mind, his originality and his industry, Ramsay's remarkable skill in manipulation, especially in constructing and managing glass apparatus, contributed in no small degree to his success in research. No man since Davy had made in chemistry discoveries of such a fundamental and far-reaching character, and no man since Davy had received such recognition alike from the scientific world and the contemporary public. Honours fell thick upon him. He was made K.C.B. in 1902, and from first to last he received honorary degrees or membership of all the most prominent Universities and Academies in the world. But those friends who enjoyed his intimacy will remember most vividly the man, his cheerfulness and good humour, his helpfulness on all occasions, his private charity and benevolence and the simplicity and purity of his life. He never forgot old friends, and in the midst of his greatest scientific triumphs he was accessible to the youngest student and to any, even a stranger, who was in need of help.

EPILOGUE

It is now about fifty years since Alexander Williamson thought it necessary to address to the Chemical Society of London a lecture on the Atomic Theory of Dalton, in which he sought to convince chemists of the physical reality of atoms as the ultimate particles to which all kinds of matter could be reduced. Ideas on this subject have advanced very far since that day. The conception of the atom, derived from antiquity and accepted by Newton, was applied by Dalton to explain some of the properties of gases, as he knew them, as well as to account for chemical combination occurring only in definite proportions. The association of atoms in groups to form molecules was an idea only slowly assimilated by the chemical world, and the word "molecule" is to be found in chemical literature only rarely before the occasion of Cannizzaro's famous exposition.

The arrangement of atoms in order within the molecule under the control of "valency" has been traced with remarkable success by the students of stereo-chemistry during the last forty to fifty years, and the means of determining the relative weights of molecules and atoms have been gradually developed and have become familiar in every chemical laboratory.

A step in an entirely new direction was taken when it was shown by Sir J. J. Thomson that the particles flying from the cathode in the experiments on electric discharge in gases under very low pressure are not molecules, as supposed by Crookes, but are much smaller than any known atoms or molecules of common matter and are, in fact, the minute bodies called "electrons" which are now recognised as atoms of negative electricity. Their mass is equal to about $\frac{1}{1800}$ of the mass of an atom of hydrogen, the lightest of known elements.

Electrons are set free in a variety of other ways, by the action of ultra-violet light on metals, and from many surfaces when heated, also especially from radio-active substances. Electrons obtained from different atoms are found to be the same, and so it appears that the atoms of each element consist of these negative particles associated with a positive nucleus which forms the greater part of the mass of the atom. It has also been discovered that the number of electrons in an atom is the same as the number representing the position of the element in a series of all the elements except argon and tellurium arranged in order of atomic weight. This is called the atomic number and follows the position of the several elements, except in the two cases mentioned, in the periodic scheme originally proposed by Mendeléeff.

In the series H = 1, He = 4, Li = 7, Gl = 9, B = 11, C = 12, etc., the order or atomic number is, 1, 2, 3, 4, 5, 6, etc., and in an atom of hydrogen there is one electron, in an atom of helium two, in an atom of lithium three electrons, and so on.

The number of molecules in a cubic centimetre of any gas can now be deduced from the results of several different methods, and hence also the dimensions of the molecule and the mass of the individual atoms.

The discovery of X-rays by Röntgen has been followed by a great extension of knowledge of many phenomena in gases. These rays in passing through a gas cause it to become a conductor of electricity, and this conductivity is due to the production of charged ions.

The study of radium, which was isolated by Madame Curie in 1899, has contributed in no small degree to the development of the subject, for the disintegration theory of Rutherford and Soddy has been fully substantiated by subsequent enquiry. We now know that the radio-active properties of this element, and of others, such as polonium and actinium, also associated with uranium ores, are due to the spontaneous breaking up of the atoms of such elements and simultaneous escape of electrically charged particles with great velocity. These particles are either electrons (β-rays) or electrically charged particles of helium (α-rays). The escape of helium from so many substances, together with the fact that the atomic weights of many known elements in the periodic table differ by four units (the atomic weight of helium) has led to the surmise that helium is a constituent of many of the common elements.

The process of radio-active change goes on through a succession of stages ending in some residual substance, an ordinary element which is no longer radio-active. In the case of radium this product is lead. By a somewhat similar series of steps from thorium lead is also produced, but the lead resulting from the one series of changes is not identical with the lead from the other, though as regards all ordinary properties they are indistinguishable. These varieties of lead and several others differ among themselves in atomic weight the value of which lies between 206 and 210. They have been called *isotopes* by Professor Soddy because they are found at the same place in the periodic scheme of the elements. Probably common lead is a mixture of two isotopes ; it has an atomic weight 207·2. But there are other elements which appear to exist in isotopic forms. Thus chlorine, which has always presented a problem in connection with its fractional atomic weight, 35·5 or thereabouts, is believed to be a mixture of two gases.

It has already been mentioned that radium in the course of its disintegration gives radiations of two kinds. One of these is produced by electrically charged particles of helium constituting α-rays, while the other is due to electrons or β-rays. The α particles moving very swiftly seem to be able to break up atoms of some of the lighter elements, such as nitrogen and oxygen, when brought into collision with them by discharging such radiation through the respective gases. According to Sir Ernest Rutherford the particles liberated from nitrogen as the result of these collisions are atoms of hydrogen. But nitrogen appears also to be capable of yielding particles of greater mass (atomic weight probably 3), carrying two charges. While it appears certain that disintegration of elemental electrically-neutral atoms can be effected by impact of the α particles from radio-active matter it is obvious that here is a wide field for further experiment, and for much speculation as to the constitution of our atmosphere and the gaseous envelopes of the sun and other heavenly bodies. In common air, for example, there must be always present not only the oxygen, nitrogen, carbon-dioxide and other gases familiar to the chemist, but various classes of ions which render the gas conductive and which may be either negative or positive and of different masses.

Among the new work which has been going on through the last few years the study of crystal structure by means of the X-rays stands out conspicuously. It is impossible to explain even the results in a few words, but for those who are interested from the chemical point of view the lecture given by Professor W. H. Bragg to the Chemical Society in February, 1916, affords the clearest exposition possible in connection with this difficult subject. Briefly and crudely stated, the end of the matter is that it is now possible to show that the atoms in crystals are set in planes separated from one another by distances comparable with wave lengths of light. The atoms are arranged in the planes in order, but there is no distribution of the atoms into molecules. In common salt, for example, the sodium and the chlorine atoms are dispersed throughout the structure in a similar manner. Each chlorine has for neighbours six sodium atoms and each sodium is surrounded by six chlorine atoms, the sodium and chlorine atoms lying alternately in any given plane. The crystal is therefore uniform in constitution throughout, and each crystal must be regarded as one molecule.

By methods such as these derived from physics a science of ultra-chemistry has been gradually evolved, and the structure of all matter and of the constituent atoms of matter is now open to the mental vision.

Though none of the chemists whose personal and experimental careers have been delineated in the foregoing pages could have foreseen either the methods employed or the conclusions reached, each one has contributed unconsciously a step or two toward the present position.

Robert Boyle for the first time gave a definition of an element as distinguished from a compound and indicated the principles by which real chemical species could be discriminated. After his time for a century the workers in the fields of chemistry were necessarily occupied in isolating from the complicated materials presented by nature the individual elements and compounds and determining the specific characters of each. The different kinds of air, mysterious as they had always been, were brought under experimental control and their relation to solid and liquid substances established. The composition of water became known and something of its real properties and functions in nature made out. By the end of the eighteenth century Black and Priestley, Scheele, Cavendish and Lavoisier had done their work, and the beginning of the nineteenth century was a period of rapid progress. The work of Davy in utilising the discovery of current electricity on the one hand and the laborious experiments of Proust, Gay-Lussac and Berzelius provided the necessary justification for the establishment of the Atomic Theory of Dalton.

The definite character of all chemical combinations, exchanges and reactions was fully realised within the first twenty years of the nineteenth century, and Faraday's discovery and enunciation of the quantitative relations between electric currents and chemical decompositions served to prove the essential identity of electrical and chemical phenomena. Avogadro's fundamental principle, though so long held in abeyance, was the means of enabling chemists to determine confidently the relative masses of molecules and atoms, and with the idea of valency, introduced by Frankland, to lead on to notions of structure within the molecule. The discovery of the non-valent elements by Rayleigh and Ramsay completed the scheme of elements conceived by Mendeléeff, while the researches of Crookes on the electric discharge in gases, and the discovery of radio-active matter by Becquerel and Curie brought the science of chemistry to the limit beyond which further progress toward a knowledge of the essential nature of matter could only be secured by the use of new methods.

Already chemical operations are largely directed by preliminary study of formulæ, and probably in generations to come, when the constitution of each element is known, the effects of bringing into contact any two elements or compounds will often be calculated beforehand and the properties of the products of any given chemical reaction predicted. Advantages in the saving of labour and material are obvious. By that time also the science of physiology will have advanced so far that the practice of medicine will be no longer empirical and, though, in the opinion of the writer, it is improbable that the origin of life and of the differences between living and dead matter will ever be finally determined, the conditions of life for the human race will be so far improved that its term will be extended generally beyond the hundred years now thought to be almost miraculous.

When that day comes it will be the duty of some future chronicler to recall the names and the labours of the chemists and the physicists to whom the advances now foreshadowed are due.

THE END

INDEX

ABOUT THE AUTHOR

SIR WILLIAM AUGUSTUS TILDEN

(August 15, 1842 – December 11, 1926)

Sir William Augustus Tilden (August 15, 1842 – December 11, 1926) was a British chemist. He discovered that isoprene could be made from turpentine. He was unable to turn this discovery into a way to make commercially viable synthetic rubber.

Life:

Educated at Bedford Modern School, Tilden obtained a B. Sc. in 1868 and a D Sc in 1871, both from the University of London. From 1872 to 1880 he was Senior Teacher of Science at Clifton College, Bristol. From 1880 to 1894 he was Professor of Chemistry at Mason College, (which later became the University of Birmingham). From 1894 to his death he was at the Royal College of Science, London, being Professor of Chemistry to 1909, Dean from 1905 to 1909, and then Emeritus Professor.

He became a Fellow of the Royal Society in 1880 and was Vice-President from 1904 to 1906. In 1908 he was awarded the Davy Medal of the Society. He was President of the Chemical Society from 1903 to 1905. The Tilden Prize was named in his memory by the Society in 1939 and has been awarded annually (now by the Royal Society of Chemistry) to three younger members since then. He held office in many other organisations, including the British Association for the Advancement of Science, the Institute of Chemistry (renamed Royal Institute of Chemistry in 1885) and the Society of Chemical Industry.

He published ***Famous Chemists: The Men and Their Work*** (George Routledge and Sons Ltd.) in 1921. His son, Philip Armstrong Tilden became a prominent architect.

Made in the USA
Monee, IL
08 November 2026

67989420R00228